BASICS
OF
ELECTROMAGNETISM

BASICS
OF
ELECTROMAGNETISM

Alimuddin Khan

ANMOL PUBLICATIONS PVT. LTD.
NEW DELHI - 110 002 (INDIA)

ANMOL PUBLICATIONS PVT. LTD.
H.O.: 4374/4B, Ansari Road, Darya Ganj,
New Delhi-110 002 (India)
Ph.: 23278000, 23261597
B.O.: No. 1015, Ist Main Road, BSK IIIrd Stage
IIIrd Phase, IIIrd Block,
Bangalore - 560 085 (India)
Visit us at: www.anmolpublications.com

Basics of Electromagnetism

ISBN 978-81-261-3577-6

PRINTED IN INDIA

Printed at Mehra Offset Press, Delhi.

Contents

Preface

This textbook is on the physics of electricity, magnetism, and electromagnetic fields and waves. The relationship between electromagnetism and life has been a source of fascination and controversy for more than 400 years. Today, interest in all facets of this relationship is at an unprecedented pitch. The body's intrinsic electromagnetic phenomena have been rediscovered, and the evidence suggests that, far from being unimportant by-products of biochemical activity as previously believed, they play a vital role in diverse physiological processes. The earth has a natural electromagnetic background, produced by the earth itself and by cosmic sources, and the age-old question as to whether this background can be detected by living organisms has now been answered in the affirmative -the earth's electromagnetic background is an important environmental factor for all living things.

The environment is now thoroughly polluted by man-made sources of electromagnetic radiation with frequencies and magnitudes never before present. Man's activities have probably changed the earth's electromagnetic background to a greater degree than they have changed any other natural physical attribute of the earth- whether the land, water, or atmosphere. The evidence now indicates that the present abnormal electromagnetic environment can constitute a health risk.This book is our attempt to synthesize the various aspects of the role of electricity and magnetism in physics, and to emphasize their underlying unity. To facilitate this, we divided primary responsibility for the major subject areas. This book treat historical factors and the regulatory role of electromagnetic energy, effects of artificial electromagnetic energy.

Author

Chapter 1

Electricity and Magnetism

"Hence have arisen some new terms among us. We say B (and bodies like circumstanced) is electrized *positively*; A, *negatively*. Or rather, B is electrized plus; A, minus. And we daily in our experiments electrize *plus* or *minus*, as we think proper."

- Benjamin Franklin

ORIGIN

The phenomena associated with electricity and magnetism was studied over most of the nineteenth century. But the knowledge that the two fields were interdependent began with the fantastic discovery by Hans Christian Orsted in the early 1820's. He learnt that magnetism is ultimately caused by moving electric charges or current, when he observed a magnetic compass needle to react to a current flowing through a wire placed near it.

Later on the simultaneous though separate discoveries made by Michael Faraday and Joseph Henry concerning electromagnetic induction in the 1830's led to the theory of James Clerk Maxwell, which united electricity, magnetism and optics into one grand theory of light: the explanation of electromagnetic waves.

Maxwell published his work Treastise on Electricity and Magnetism, in which he showed that four fundamental mathematical equations described the entire known electric and magnetic phenomenon. The first equation is Gauss's law for electricity, which states that positive and negative charges create magnetic fields; Gauss's law for magnetism states that currents create magnetic field, which have associated north and south poles, but single poles (monopoles) do not exist; Ampere's law states that time varying magnetic fields induce time varying electric fields; and faraday's law of induction states that time varying electric fields create time varying magnetic fields. Additionally, Maxwell's equation predicted the existence of combined, changing electric and magnetic fields in the form of waves that traveled with the speed of light i.e. electromagnetic waves. He speculated that accelerated charges ultimately create these electromagnetic waves, that they should exist over a wide range of frequencies and wavelengths, that they traveled at the speed of light in a vacuum, and that they exhibited all the optical properties of visible light, such as reflection, refraction and diffraction.e

Heinrich Rudolf Hertz in 1887 verified Maxwell's theory experimentally ten years after his death. Hertz built an induction coil device, which was essentially a step up transformer whose high output voltage caused, sparks to jump back and forth across an air gap between two metal plates. One wire, bent so that it too had an air gap between its ends, was placed near another wire. Hertz noticed sparks jumping across the ends of this wire at the same frequency as the induction coil's sparks. He concluded that electromagnetic waves propagated through air from the coil to the bent wire. These waves proved to be radio waves of about 1 meter in wavelength. He demonstrated that the waves exhibited all the usual properties of light; namely, they reflected, focused on parabolic mirrors, and refracted through glass. He caused them to interfere, setting up a standing wave pattern that enabled him to calculate their speed to be the speed of light. Later experiments demonstrated that a wide range of electromagnetic wavelengths and frequencies exist and led to the technologies of radio, television, radar and myriad other technologies important to society.

THEORY

Many natural phenomena exhibit wavelike behaviour. Water waves, earthquake waves, and sound waves all require a medium or substance through which to propagate. These are examples of mechanical waves. Light can also be described as waves- waves of changing electric and magnetic fields that propagate outward from their sources. These electromagnetic waves however do not require a medium. They propagate at 3,000,000,00. meters per second through vacuum. Electromagnetic waves are transverse waves. In simpler terms, the changing electric and magnetic fields oscillate perpendicular to each other and to the direction of the propagating waves.

The best source of electromagnetic waves is accelerated waves. An accelerated charge is one that is increasing or decreasing its speed or changing its direction of motion or both. Let us imagine two charges at rest in the vicinity of each other. They are immersed in each others electric force field. If one charge suddenly begins to oscillate up and down, the second charge experiences the change in the field of the first charge after some very small finite time elapses. The oscillating charge was accelerated. The moving charge's electric fields change, as do their magnetic fields. These changing electric and magnetic fields generate each other through Faraday's law of induction and Ampere's law. These changing fields dissociate from the oscillating charge and propagate out into space at the speed of light.

All periodic waves, whether they are electromagnetic or mechanical, are characterized by such properties as wave length, frequency, and speed. For electromagnetic waves, wavelength measures the distance between the successive pulses of electric or magnetic fields. A waves' frequency represents how many wave pulses pass by a given point each second and is measured in cycles per second or waves per second and is measured in cycles per second or waves per second. One wave per second is called one Hertz. Electromagnetic waves travel at the speed of light in vacuum, but they travel more slowly when they pass through various media such as air, glass, and water. A relationship among frequency, wavelength and speed exists for electromagnetic waves; the product of frequency and wavelength equals the speed of light. Thus, wavelength and frequency are inversely related. The longer the frequency lower is the wavelength and vice versa.

An entire spectrum of electromagnetic waves exists, which ranges from very low frequency wavelength (power waves) to very high wavelength (gamma rays). All wavelengths are collectively referred to as electromagnetic wavelengths and not merely the narrow range of wavelengths and frequencies identified as visible light.

The wave nature of light describes many aspects of its behaviour. Nevertheless, radiation also has its particle like characteristics. Rather than infinite or nearly infinite series of electromagnetic waves emanating from some accelerated charge, light also appears to come in particle –like bursts of energy. These individual bursts of energy or quanta are called photons. Each photon possesses an amount of energy that directly depends on the frequency of the associated electromagnetic wave. Doubling the frequency of the photon of radiation doubles its energy. Thus, all types of electromagnetic waves, photons of power waves possess the least energy and gamma-ray photons possess the greatest energy.

Since life on earth is bathed constantly in all forms of electromagnetic radiation, scientists must be aware of the potential risks, as well as benefits of exposures to electromagnetic waves.

ELECTRIC CHARGE A PROPERTY OF MATTER

As with all other physical phenomena associated with motion, the "explanation" of electrical phenomena starts only with observations. The experiments are more difficult to replicate and harder to explain than those relevant for motion.

Still, if you compare enough observations involving *static* electrical phenomena (e.g. dragging your feet across a carpet and touching a door handle, rubbing plastic piping with cotten, etc.) you can, as Benjamin Franklin did, arrive at the following conclusion: Matter can have a physical property which we call electrical charge

Electrical charge, like mass, is an intrinsic property of matter. It shares many of the properties of mass in fact. The most important ones that we have discovered over the ages are that

1. Charges create and are subject to electrical forces.
2. Charges can neither be created nor destroyed, i.e. they are intrinsic to matter.

In addition, there are some things that charges do not share in common with mass:

1. Charges come in two types: designated as positive and negative.
2. Charge is conserved as stated above, but it can readily be transferred from one object to another as long as the object is not a fundamental particle (e.g. an electron).

That last caveat is important. It turns out that, ultimately, charge and mass do share the property of being *intrinsic* to matter, i.e. elementary particles have a definite mass and a {\it fixed} charge. Thus, the fact that we can transfer electrical charges between objects is explained by stating that this takes place by the actual exchange of charged *particles*. Thus, the charges involved are not created or destroyed anywhere, just transferred from one object to another by the exchange of small (essentially invisible) charged particles (which we now know are usually, but not alwaygs, electrons).

As neat as the above explanation is, it still does not explain all electrical observations you can make though. For example, two elementary particles (e.g. an electron and a proton) exert forces on each other without the exchange of anything "charged" between them. We can also see that macroscopic objects which are "neutral" can be subject to electrical forces without making any physical connection between them and any other object (this is referred to as induced charging). Hence we have to infer that "something" permeates the space between electrically charged objects to exert a force on them. We explain this "something" as being the electric field. Before describing it's properties, let's first work backwards, or operationally, i.e. let's say what the field does and let that serve as our definition of what it is.

COULOMB'S LAW

The force between charged particles was described on the basis of sensitive experiments by Charles Augustin Coulomb in a formula known as Coulomb's Law. Assume two charges, 1 and 2, are separated by a distance r_{12}. Since charges come in two types, we can have two cases: the charges are the same "sign" (i.e. both positive or both negative) or they are opposite (i.e. one is positive and the other negative). For either of these two cases, Coulomb found that the magnitude of the force between the particles was

$$F = \frac{k|q_1||q_2|}{r_{12}^2}$$

where the magnitude of the charges are expressed in units of Coulombs (C) in the SI system of units. The proportionality constant k has an experimentally determined value of $9.0 \times 10^9\ N \cdot m^2/C^2$. To complete the picture, we have to define the direction of the force and define what it acts on. Experimentally, we observe for our two cases (i.e. both charges have the same sign or the charges have opposite signs) that the forces acting look as follows:

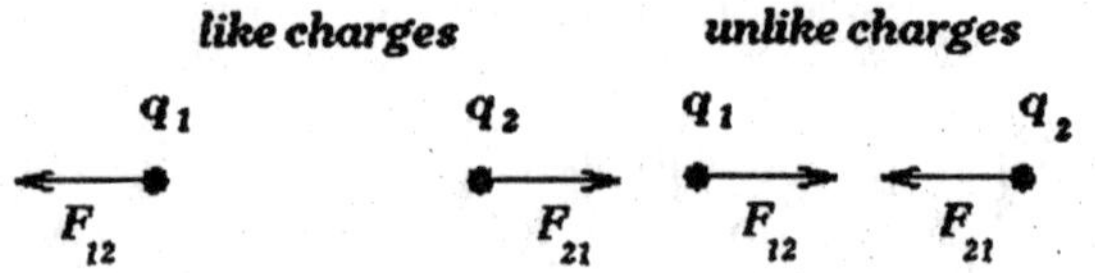

Thus, to complete our definitions of the forces produced, we go to the vector expression of Coulomb's Law.

$$F_{21} = \frac{kq_1q_2r_{12}}{r_{12}^3}$$

This is the force on charge q_2 due to charge q_1. To get the proper direction, we need to note two things.

i. The quantities q_1 and q_2 now have signs which reflect the sign of the charges they represent, i.e. they can be either positive or negative.

ii. Note that the vector direction is given by the unit vector r_{12}/r_{12}. The vector r_{12} extends from q_1 to q_2, i.e. *from* the charge providing the force *to* the charge that the force acts on.

Let's consider an application of Coulomb's Law. Since the elementary case of a single charge is already explained, let's consider three stationary charges. We will not specify the magnitudes or even the signs of the charges to emphasize that the approach to the problem of calculating the forces is independent of these actual values. The answer certainly depends on having this information, but the solution depends only on the mathematical nature of electrical forces as expressed through Coulomb's Law.

Consider three electric charges arranged as a triangle. Find the net force acting on charge q_3.

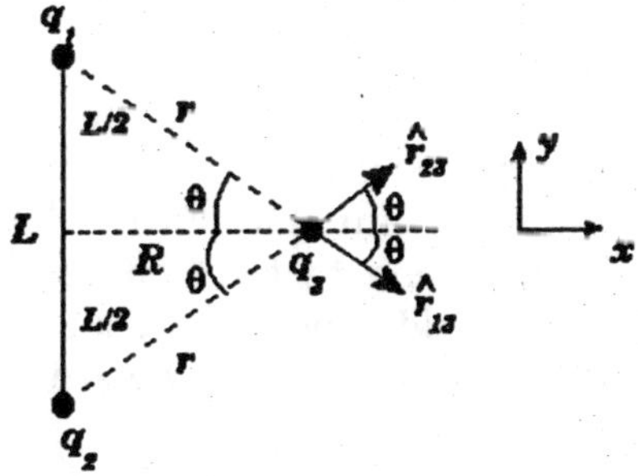

The forces exerted by two charges, q_1 and q_2 on a third charge, q_3, must lie parallel or antiparallel to the directions given by the unit vectors $\hat{r}_{13}$ and $\hat{r}_{23}$, respectively. Charge q_1 must exert a force on q_3 which lies either along (q_1 and

q_3 have the same sign) or anti-parallel (q_1 and q_3 have opposite signs) to the vector r_{13}. The force of charge q_2 on q_3 must lie either along (q_2 and q_3 have the same sign) or anti-parallel (q_2 and q_3 have opposite signs) to the vector r_{23}. This will automatically be reflected in our answer if we adopt the convention of letting the direction we choose be from the charge causing the force *toward* the charge which experiences the force.

ELECTRIC FIELDS AND ELECTRIC FORCES

Although we could go on to describe experimental observations of forces determined from endless combinations of electrical charges with various numbers, signs, magnitudes of charges, distances, and orientations, let's take an approach that will prove enormously useful. That is to introduce a mechanism for how the force is created. Since charge and mass are analogous, let's assume that the mechanism for how mass "creates" gravitational force is copied by charge for "creating" electrical forces. We call this mechanism the electric field. We have to make the properties of electric fields consistent with the experimentally observed properties of electrical forces. A simple scheme for doing this is to make the field and the force proportional to one another, hence we define the electric field, E, as E = F/q

That is to say, the force on a charge q which is in the presence of an electric field E is F = qE. Note that the bold letters used indicate that E and F are vectors. The relationship we have used makes it easy to define the properties of E in terms of the point charges that produce them. Since we have already seen how one point charge (the charge, say on an elementary particle) can exert a force on a second, we should state that the electric field produced by a positive (negative) point charge goes radially outward (inward) in all directions from (toward) the charge.

ELECTRIC FIELDS

The field lines are a (rather gross) characterization of the actual electric field. The field extends radially, in all directions in 3-dimensional space, from the charge. If the charge is positive, the field is said to point away from the charge. If the charge is negative, the field points in towards the charge. The field lines give a general idea of the direction, but it is important to remember that the actual field in 3 dimensions extends to every point in space out to infinity if there are no other charges or matter. The weird shape of the field for two or more charges comes about strictly for geometrical reasons. The electric force, we find experimentally, is both conservative and obeys the principal of superposition (just as gravitational forces do). Therefore, electric fields must also obey these rules. The superposition property means that electric fields from two different sources add *vectorially* to produce a *net* field. Let's look at two point charges as in lecture 1. The four possible combinations of positive or negative charges are shown.

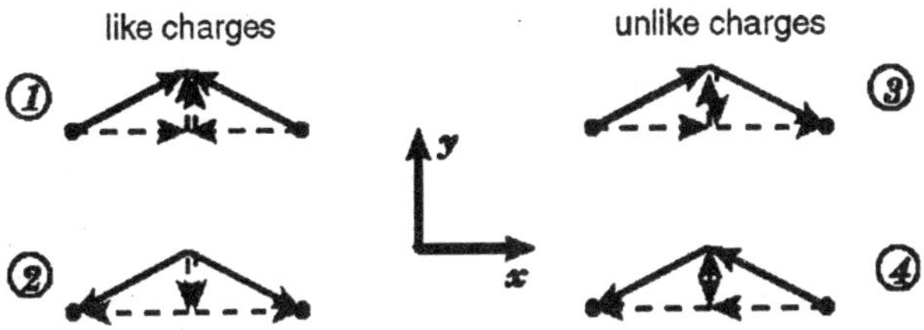

Fig. Vector components for electric fields from two charges.

Note that we have chosen the same position (relative to the charges) to calculate the net electric field in each of the four cases. For cases 1 and 2 (both charges the same sign), we note that the horizontal or x components of the fields tend to cancel, leaving the y components to add. This will be true for any point which is close to the line which is parallel to y and located at an x position close to the midpoint between the charges. The field lines indicate that the field is nearly vertical for points close to the midpoint line as we have described it. For cases 3 and 4 (unlike charges), the position we have chosen for calculating the net electric field tends to have the y components canceling and the x components adding. Hence any point near the midpoint line for unlike charges will have electric fields pointing nearly horizontally. Right along the midpoint line, the y component is exactly zero (assuming the two charges are equal magnitude). To draw the field lines for two charges as in the text or in the Java applet, we need to find the vector sum of the electric fields extending radially from or toward each charge (depending on whether the charge is positive or negative) at each point in space, then, for some small subset of points, find the direction of the net electric field. The field lines in the book or in the applet show the general direction of the electric field for various positions around the charges. Thus, the shape looks complex, but it is arrived at by doing the vector sum of radial field lines from each charge, so nothing complex is going on.

To find the force on one charge due to other charges then, we note that:

1. No point particle is subject to forces from electric fields produced by the charge of the particle itself
2. Superposition says that the net force on a charged point particle due to the presence of other charged point particles can be derived by first finding the electric field due to the other particles, then using qE = F.

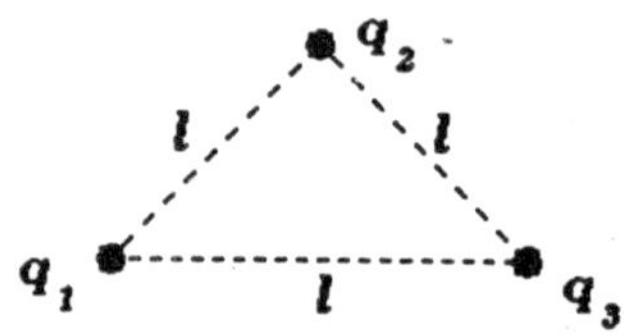

Fig. Three charges fixed in a triangular pattern.

Let's try this out by find the net force on the lower left particle in the arrangement shown below.

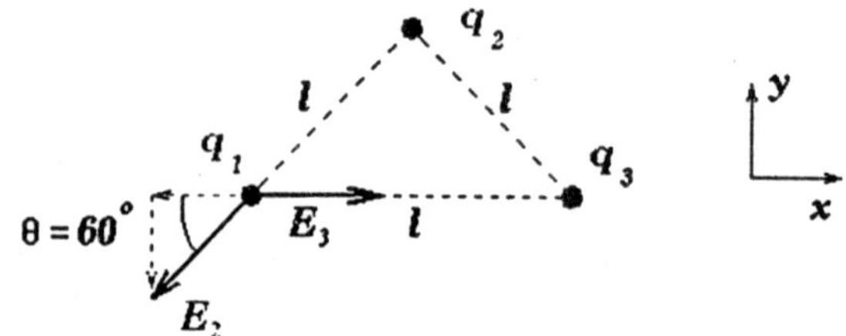

Fig. Electric fields on charge q_1.

Assume that the red particles are positively charged and the blue particle is negatively charged. Since the particles are along the vertices of an equilateral triangle, the relative angles with respect to a convenient x-y axis is easy to determine. We observe the following for the contributions to the electric field at the lower left corner:

Note that only the fields due to q_2 and q_3 have to be considered as the force on q_1 is determined only by *external* charges (for the case of point particles).

CONTINUOUS CHARGE DISTRIBUTIONS

Naturally, we cannot be content with considering only the fields or forces between point particles. Real objects, however, can be considered as being made up of a large number of very small pieces which *approximate* points. If we understand how charge is distributed over an object and if the object and its charge distribution have a high degree of symmetry, then we can easily calculate the electric field due to that object, and hence the force it can produce on other charged objects, by the usual techniques of integral calculus.As an example of the above statement, consider a charged line segment of length l and charge +Q. Here, we have not a point, but an infinitely thin line of charge. Without any other information to go on, we assume that +Q is uniformly distributed throughout the line. If we wish to know what the electric field is at a point a distance h above the midpoint of the line, how do we calculate this?

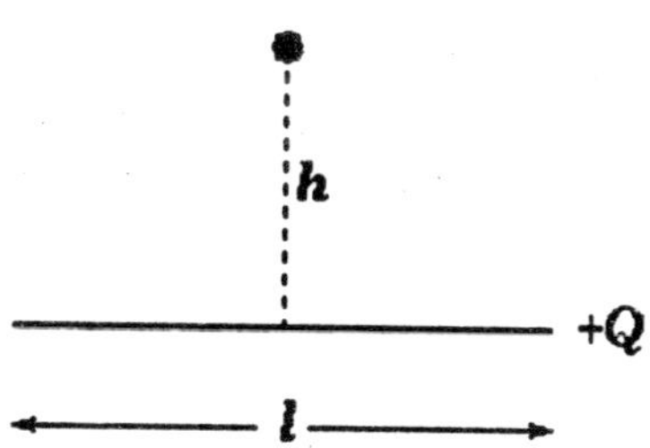

Fig. A linearly distributed charge.

We approach this problem by breaking the line into small segments of infinitesimal length. The lengths are small enough that we can assume that each little segment can be considered as a point (later we can make a self-consistent check of this assumption). Before we go through any calculation though, we consider how symmetry can reduce the amount of work we need to do. In this case, if we consider the result of two infinitesimal segments, one on either side of the midpoint line as shown below, we note that, as in the previous case of two point charges, the horizontal components of the electric fields tend to cancel in making the net field.

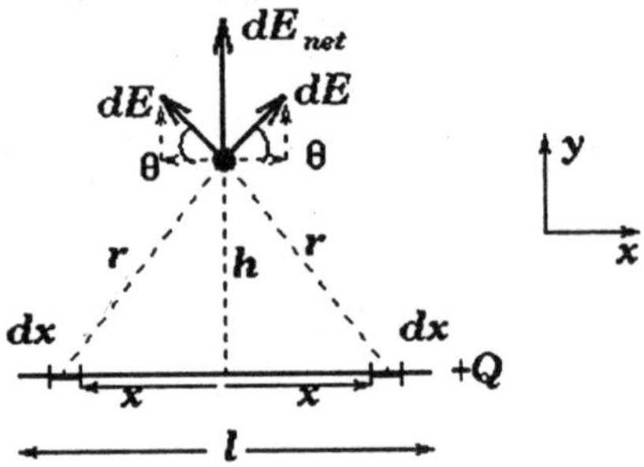

Fig. Calculating the electric field due to a charged line segment.

Thus, we note that we only need to keep track of the vertical components for the line segments as the horizontal ones will cancel out. Furthermore, we can get away with only adding up the contributions to the net electric field from one half of the line segment and multiplying by two to take care of the contribution of the other half. This is not much work in this case, but it's nice to be clever when you're sure you're right. In this case, the integration can proceed once we know how to write down dE. To get that, we need to consider the charge of the infinitesimal line segment. If we assume that the charge is uniformly distributed, then the charge of each infinitesimal segment must be

$$dQ = Q\frac{dx}{1}$$

Another way of saying this is to consider the *linear charge density* as being Q/*l*. Then the charge of any length, dx, of the line segment is (Q/*l*) dx. The vertical component of the electric field due to any little length, dx, is therefore

$$dE\frac{\sin}{2} = \frac{kQdx}{lr^2}\frac{h}{r}$$

where $r^2 = (x^2 + h^2)$. Now we can set up the integration and finish the problem. For convenience, let's place the origin of our coordinate system at the midpoint of the line segment. Maple makes quick work of the integral.

Let's go back and check that we have self-consistency with our original assumption about very short line segments looking like point charges. If we consider the above answer, we can ask the question: what does the electric field look like if l, the length of the line segment, is itself very small compared to h. In other words, if h is big, then the point where we want to calculate the field is very far away from the segment and we expect that it will be difficult to tell the difference between the field due to a point charge and the field due to a small line segment. On the other hand, if h is not very large but l is very small, again, we assume the field looks like that due to a point charge. Thus, the formula for the net electric field does go to that of a point charge (kQ/h^2).

ELECTRIC FIELDS OF FINITE OBJECTS

We can define the electric field for any number of distributions now that we know how to apply Coulomb's Law in combination with integral calculus. One of the hallmarks of truly "learning" a subject though is that you first learn "what" neat things you can do, then you learn when not to do them. In other words, it's much more important to understand the importance of symmetry in determining electric fields than it is to memorize the formulas for any particular charge distribution. Symmetry reduces the amount of effort required to get from charge distributions to E fields to the point where the memorization is actually more work than deriving the answer when you need it. A couple of examples are in order:

ELECTRIC FIELD FOR A RING OF CHARGE

The ring of charge below has a charge +q uniformly spread around a ring of radius R as shown below.

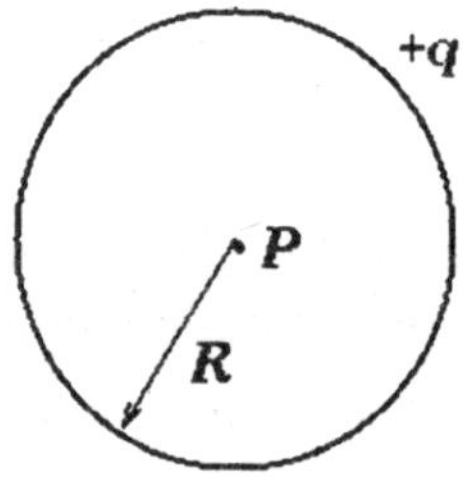

What is the electric field direction and magnitude at the center of the ring?

In this case no calculation is necessary. If we break the ring into infinitesimal pairs, each with charge dq and arc-length ds, we see that the electric fields from the pairs cancel as long as the members of the pairs are chosen so as to be on opposite sides of the ring as shown.

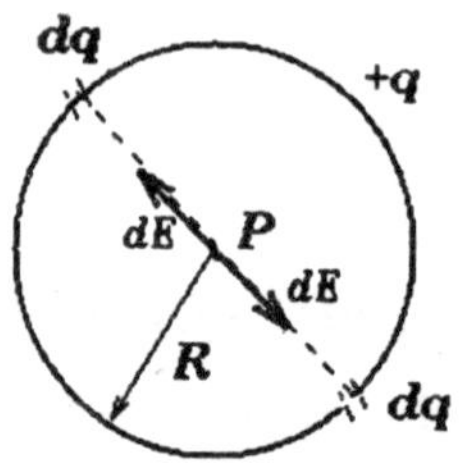

Fig. Electric field at the centre of a uniform ring of charge +q.

That is to say, the electric field is zero since each infinitesimal dq we pick around the ring has its field canceled by another dq diametrically opposite it.

ELECTRIC FIELD FOR A DIPOLE RING CHARGE

Suppose the ring above has the top half possessing a positive charge +q and a negative charge -q on the bottom half.

Now we need to do some work. First note that there are *two* symmetries at work here.

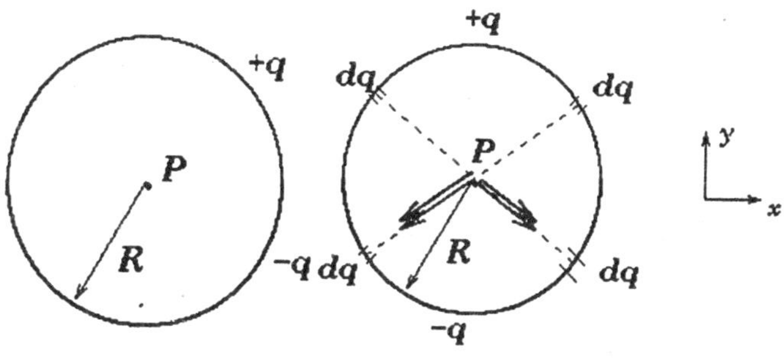

We see that for any pair of infinitesimal arc-lengths on the positive half-ring we can choose the elements so that the horizontal (i.e. x) components cancel while the vertical (i.e. y) components add. If we choose pairs of infinitesimal arc-lengths from the negative half-ring in the same way, we see that their field components also add vertically and cancel horizontally.

Thus, we can say on the basis of symmetry alone that the *net* electric field points downward at point P. With the direction known it is now just a question of calculating the magnitude of the net E field. For each infinitesimal arc-length, we have $dE = k\, dq/R^2$ as the magnitude of E field produced at the centre. To give each dq a location, we can use the angle w.r.t. the vertical (note that we could use the angle w.r.t. the horizontal just as easily). Then, this angle defines the vertical component of dE at point P.

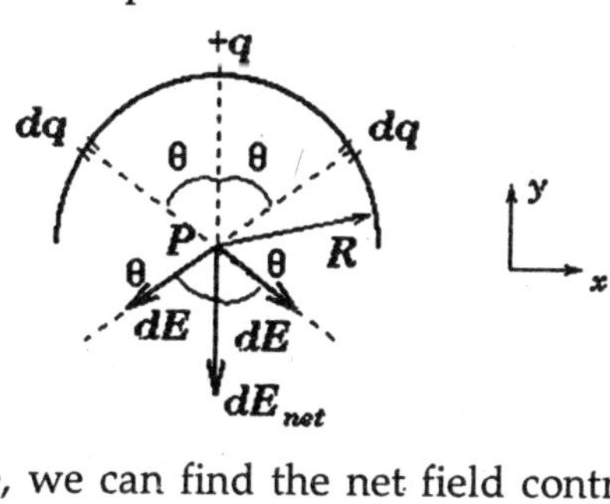

Therefore, we can find the net field contribution of the pair of infinitesimal arc-lengths as

$$dE_{net} = dE_y + dE_y = \frac{dq\cos\theta}{R^2}$$

Now we take advantage of one of our symmetries to only calculate the vertical components since the horizontal components all add. We can take advantage of the second symmetry when we are done calculating the field due to the top half-ring since the bottom must contribute the same field at point P. Thus, we do the calculation for the top half then multiply by two to get the answer for the entire charge distribution (top half-ring positive and bottom half-ring negative). To do the calculation for the top half-ring, we need to parameterize dq in terms of theta. That's easy enough in that the linear charge density of the top half-ring is q/(p*R) since p*R is the circumference of a half-ring. The arc-length of dq is R*dq. So, now we are set to do our integration to find the net field due to the top half-ring. Note that, since we are doing pairs of dq's, we only need to integrate from 0 to λ/2 (i.e. a quarter-circle). Also note that we will use λ for the linear charge density.

ELECTRIC FIELD FOR A RECTANGULAR SHEET OF CHARGE

Find the direction and magnitude of the electric field for a sheet of charge –Q uniformly spread over an infinitely thin rectangular sheet of length *l* and width w, at a point P located a distance h above the midpoint of the sheet.

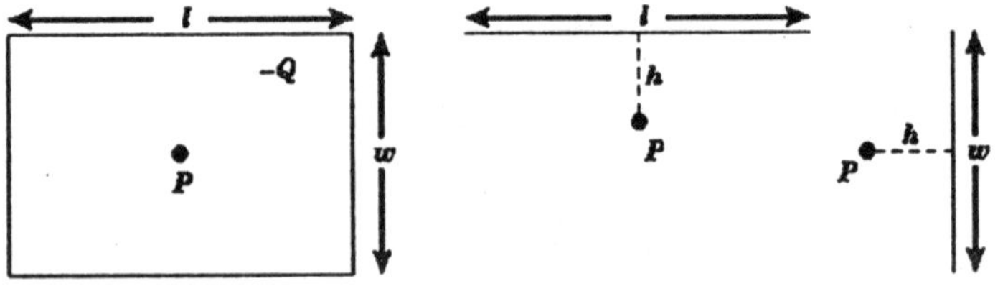

front view ***top view*** ***side view***

An infinitesimally thin sheet of charge –dQ with length *l* and width w. The electric field is evaluated at point P a distance h above the midpoint of the sheet.

ELECTRIC FIELD FOR A RECTANGULAR SHEET OF CHARGE

We could do this problem by breaking the sheet of charge into tiny rectangles of length dx and width dy, treat these infinitesimal rectangles as point charges, find the electric field contribution at point P for each rectangle, then do a two-dimensional integral to find the net electric field at point P. This would be quite tedious. We can instead note that there are symmetries here which allow us to exploit a solution to a previous problem. First, note that our result for the electric field at a point located a distance h from the midpoint of a line of charge q with length l is

$$E_{line} = h\frac{2kq}{l^2 + 4h^2}$$

We make use of it now by noting that the result is true for *any* point a distance h from the midpoint of the line of charge no matter what it's orientation. This is called cylindrical symmetry. The direction of the field is radially inward, i.e. toward the line of charge, if the charge is negative, but the magnitude is unaffected. That means we can use this result for the current problem since all we need do is to assume that our sheet of charge is composed of an infinite number of infinitesimal lines of charge, each with length l and width dy.

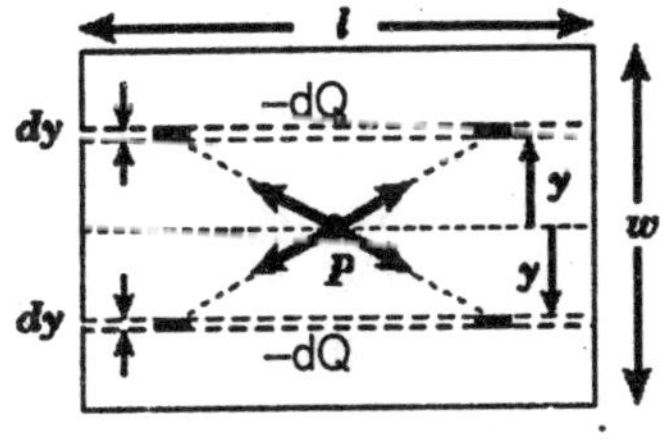

Front view

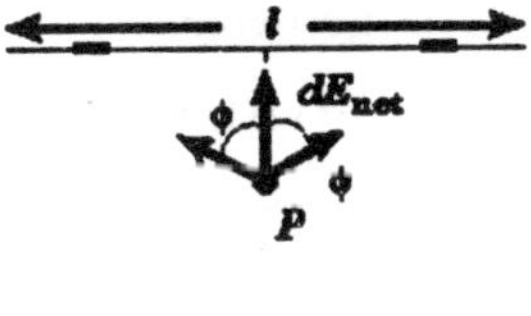

Top view

A sheet of charge -Q of length l and width w broken into lines of length l and width dy.

Note that we show only two of these lines, both a distance y from the midpoint, to point out the symmetry that proves that the net electric field of the two is directed perpendicular to the plane and toward it from point P.

We define the position of each line as distance y from the centre of the sheet and let each have a charge dQ. Then the magnitude of the electric field at point P due to each line is

$$dE = h\frac{2kdQ}{l^2 + 4h^2}$$

The direction of this field contribution is towards the line. If we consider *pairs* of lines, each having charge dQ, and located y above or below the centre of the sheet, then we can exploit cylindrical symmetry again: each rectangle contributes a field whose component parallel to the sheet cancels the parallel component of the other rectangle. We are left with the perpendicular components of the electric field due to both rectangles adding. Therefore, for two rectangles, a distance y above/below the centre of the sheet, the net electric field has magnitude where f represents the angle of the electric field vector with respect to a line perpendicular to the plane of the charge sheet. To integrate over all rectangular charges dQ, we first need to note that

$$dQ = \frac{-Q}{w}dy$$

Then the net electric field when we integrate over all pairs of rectangles is directed from point P towards the midpoint of the charged sheet.

DEFINING GAUSS' LAW

To make more progress on understanding the role of electric fields in explaining electrical phenomena, we need to specify a way in which the nature of the electric field produced from a given charge distribution manifests itself in space. Hopefully you have seen from even the few examples we have done with Coulomb's Law just how important the role of symmetry can be in reducing the complexity inherent in the problem of this "specification". Since Coulomb's Law works for point charges and forces us to reduce every finite distribution of charge into points, it clearly does not make very powerful use of symmetry (the only symmetry available is that due to a point charge and the ability to reduce the task of the integration by choosing integration strategies that maximize the use of symmetry to reduce the work of evaluating the integral).

An indirect approach created by Karl Friedrich Gauss does intrinsically bring in symmetry in the formulation of the method for explaining how an electric field is generated from a charge distribution. It starts by noting that a proper description of the electric field invokes the notion of a vector field. The vector field simply specifies the direction and magnitude of vectors at any particular point in space you care to look at. Pictorially, we can imagine something like the picture below:

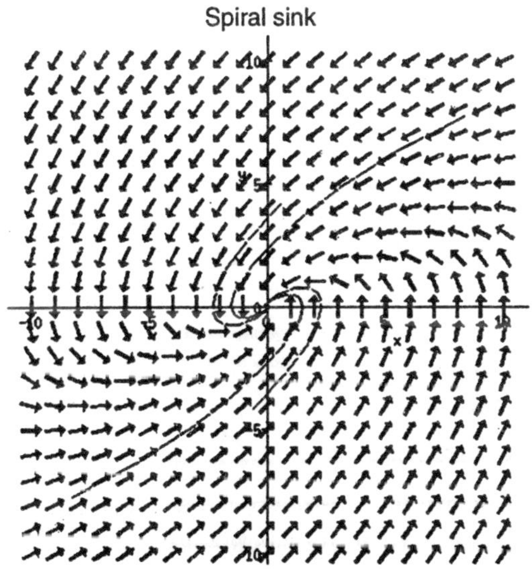

Fig. A sink.

The arrows here depict the direction and magnitude of something that can be depicted by vectors at each point in space. This probably reminds you of something like water flowing down a sink. In fact, it could represent a number of situations in which something is coupled to an event which leads to a progression in time which tends towards zero. The vector field above could represent the velocity, at various points, in a pool of water flowing down a drain located at the origin of the coordinate system or any number of other things, e.g. the inflow of dust circling around a nascent star. This tells us of the tremendous power of vectors for description. Any kind of information can be represented in vectorial form if we make the correct associations. Vector fields can also be used to represent electric fields. The field lines we draw are, in fact, a crude representation of a vector field. Gauss realised that a more useful quantitative representation of how things *flow* throughout space could be realised by being more precise about how a vector field is defined. We can start by being more precise about where the vectors should be drawn from, i.e. use a more definitive way of specifying how to draw the vector field so as to maximize the information we can get from it. We can start by being more orderly as to where we draw the vectors. If, for example, we start from a given surface, we can state what the vectors look like in defining the flow of something from a specified *set* of points which can be used to define an initial position. The flow of something can be defined at the surface A and we know we have a definitive description of the flow at the position of A. Without this convention, saying "where" the flow is at any given time is somewhat ambiguous.

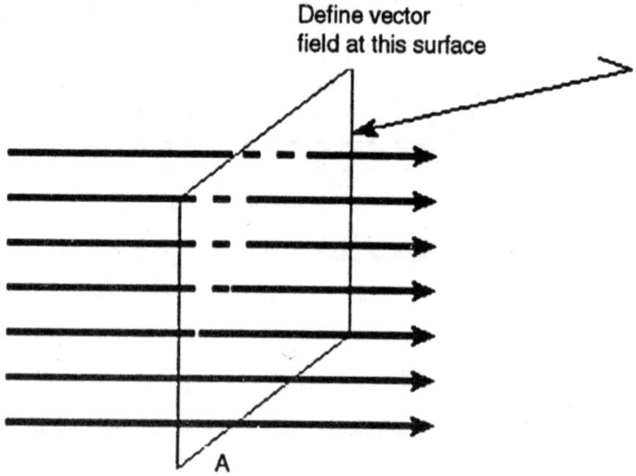

Fig. The vector field defines the flux through surface A.

Although it might not appear as though this type of "defining" might be useful, it already quite powerful for talking about physical observation. For example, if you've ever noticed water flowing from a faucet set open to give a relatively small flow of water, you might have noticed that the flow narrows as the water falls away from the faucet opening.

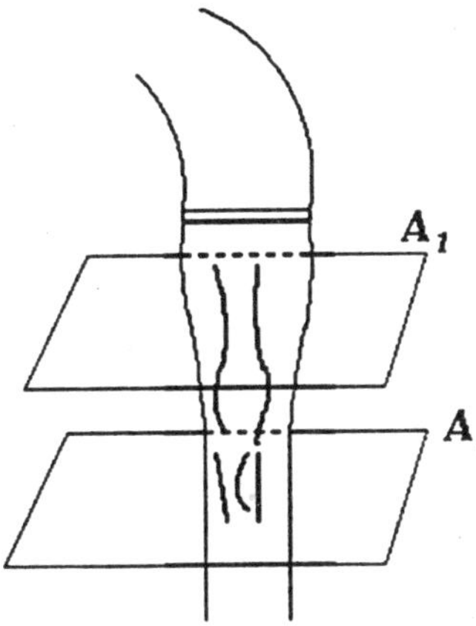

The flux through the two areas A_1 and A_2 must be the same.

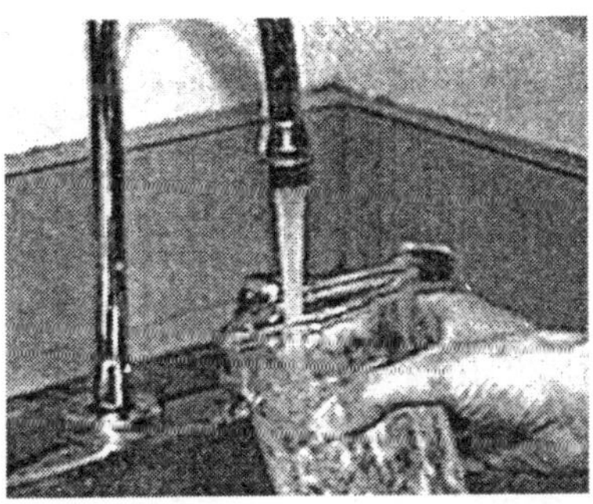

If we look at the *flux* of water, defined as the amount or mass of water flowing across an area, A_1, per unit of time, then we know that the same mass of water per unit time must flow across A_2, placed at a lower position. However, since gravity causes the speed of the water to increase as it falls, the cross-sectional area of the water, that is to say, the size of the stream going through A_1 must be larger than the size of the stream going through A_2 since the volume of water (and hence the total mass since the water has constant density) per unit time crossing A_1 or A_2 is given by (cross-sectional area) × (velocity). Hence, since velocity goes up, cross-sectional area must go down. Note that water reaches its terminal velocity quite quickly so that the stream reaches a constant cross-sectional area after just a small distance of travel. To be precise about measuring flux, we should define that the flow of something through a surface is dependent on the angle between the direction of flow and the surface. Since any vector can be resolved into any set of orthogonal components, we should state that the flux is only proportional to the component of the flow vector which is *perpendicular* to the surface. This turns out to be such a useful case that physicists generally define area as a vector! The magnitude of this vector is what we normally think of as the size of the area and the direction is always perpendicular to the surface. Since you always have two directions in which you can be perpendicular, the convention is that the direction is chosen so as to point away from any volume which is *enclosed* by the surface whose area we are interested in. In other words, physics generally is interested in areas which confine (in fact, define) some volume. Pointing away from the volume distinguishes inside the volume from outside the volume. The flux is then defined by the dot product of a vector representing the flow and this vector. We can state this relationship mathematically as *flux* = $v \cdot A$ where v is the vector describing the flow and A is a vector perpendicular to the purple area (and directed along the flow by convention) with magnitude equal to the area covered by the flow. While v and A are parallel, the flux is indeed a maximum (the dot product is a maximum when the two vectors of the product are parallel) and equal to zero when v and A are perpendicular (in that case the purple area is parallel to the flow).

SPHERICAL SYMMETRY

The first practical application of these ideas of flux is to a mystery first explained by Newton. In his theory of Universal Gravitation, Newton speculated that his force formula for gravitational attraction between two *point* masses, m_1 and m_2, should work equally well for *spherical* masses m_1 and m_2. If we replace a spherical mass distribution by a point at the centre of the spherical distribution with the same total mass, the gravitational force looks exactly the same to the universe outside the sphere. Gauss would say that this "coincidence" works because points and spheres possess the same symmetry (spherical symmetry). Imagine a point charge, q_1. The charge emits an electric field in all directions. No direction is preferred. If we replace the electric charge by a spherical distribution of charge q_1 with radius R, the same condition holds. The sphere looks the same from any angle so the electric field it produces looks the same from any direction. Put any other charge further away from q_1 than R and it cannot tell the difference between the point charge and the spherical charge distribution. Gauss takes advantage of this to relate his electric flux formula for the spherical charge distribution to Coulomb's Law for a point charge. Gauss states that electric flux is $\Phi_E = E \cdot dA$ and that this is proportional to the electric charge. Some comments are in order: first, note that we use an integral to define the flux. This accounts for the fact that we generally need to define the flux through an area which has to be broken into infinitesimally small pieces. The flux through each piece is evaluated and the total determined via integration.

Gauss' Law is the following: $F_E.q_{enc}$ for an area which *encloses* (hence the circle through the integral) a charge q, the electric flux through this area is proportional to the charge enclosed. Assuming this to be right for the moment, we can evaluate this for a positive spherical charge distribution of radius R. Since the electric flux extends outward, if we choose the surface of the sphere as the surface from which to evaluate the flux, then E is perpendicular to the surface at each point (since there are no preferred directions) then E · dA is just E dA since the vectors are parallel at every point on the spherical surface. Also, since the sphere is assumed to be uniform, the E field has equal magnitude along this surface, hence the integral of E.4dA is just EA = E(4pR^2). Thus, if the proportionality constant is called e_0, then

$$q = \varepsilon_0 E(4\pi R^2)\ E = \frac{q}{4\pi\varepsilon_0 R^2}$$

Clearly, to be consistent with the result expected from Coulomb's Law we need

$k = 1/4\pi\varepsilon_0, \varepsilon_0 = 8.854 \times 10^{-12}\ C^2/N \cdot m^2$.

Thus, Gauss' Law is expressed as

$$E \cdot dA = \frac{q_{enc}}{\varepsilon_0}$$

with the subscript enc meaning the charge enclosed in the volume contained by the surface for which the flux is defined.

APPLICATIONS OF GAUSS' LAW

Let's find the electric field due to a line charge. As we have done this before, much of the setup of the problem is already done Let's make things a bit tougher by considering the field due to an *infinitely* long line of charge as opposed to the one of finite length which we did before. It's clear here that it's impossible for us to talk about a finite amount of charge stretched over an infinitely long distance. Instead, we state that the line has a constant linear charge density, l.

Realistically, all line charges are finite, but, first, we have done the finite length problem explicitly, and second, a line of charge which is quite long compared to the distance from it at which we would like to know the E field is not an uncommon problem. We can deal with the approximate answer as easily as with the true solution and compare the differences. Physicists will often engage in thoughts of this kind in theoretical research as it is important to know where our ideas "break down". Those ideas which can hold up under the most extreme extrapolations without delivering clearly nonsensical results indicate something deep about our understanding. The above being said, we still need to do the calculation. Consider the figure which shows a view of the line charge and a point P a distance h away from it. We wish to find the electric field at point P. To set up the integral, we do, as before, the trick of taking infinitesimally small line segments of charge in pairs so that their horizontal components cancel and the vertical (i.e. radial) components add.

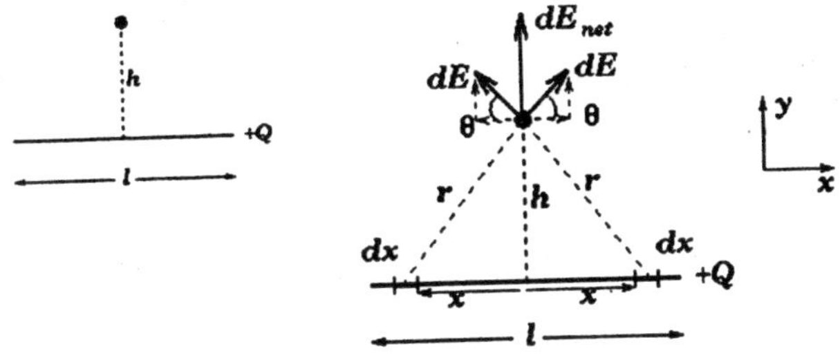

Fig. Calculation of the electric field at the midpoint of a line charge of length *l*.

Hence we only need to change the definition of dq, the charge on an infinitesimal segment with length dx, otherwise our approach for the finite line charge is unchanged for the infinite length case. Now, $dq = \lambda dx$, so, as with the finite length charge, we use the angle, q with respect to the vertical to identify the radial component, r, as the distance from the infinitesimal charge to point P, i.e. and thus We calculate the magnitude of E as follows. $E_{net} = \lambda/2\lambda\varepsilon_0 h$.

We can also approach this problem using Gauss' Law.

The Gaussian surface should be a closed, hollow can. The sides of the can are perpendicular to the E field (since we just determined previously that the E field is radial only, i.e. there are no components perpendicular to the radius). Hence E·dA is just E dA since the electric field also only depends on h as we just showed with the explicit integration done earlier. If we make the Gaussian can of finite length (we should since we will need to find the amount of enclosed charge), say, L as in figure, then the charge enclosed is l L. The calculation of the E field proceeds in this way: $E_{net} = \lambda/2\pi\varepsilon_0 h$

We get the same result, but with somewhat less work. Gauss' Law is useful for calculation of the electric field for situations which have a high degree of symmetry.

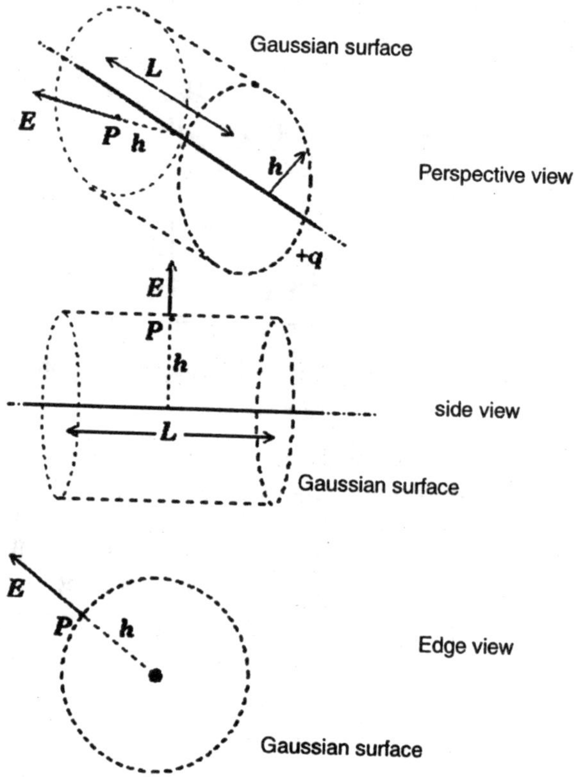

Fig. Gaussian Surface with Cylindrical Symmetry Surrounding a Line Charge.

In fact, for freshman physics, we can generally only apply it in cases in which one of the following symmetries are exhibited:

- Spherical symmetry (point charges, spherical charge distributions)
- Cylindrical symmetry (infinitely long lines of charge, infinitely long cylindrical charge distributions)
- Planar symmetry (infinite sheets or blocks of charge)

We have discussed examples of the first two of these. We can proceed now with the latter.

PLANAR SYMMETRY

In the case of charge distributions which exhibit planar symmetry, i.e. that have a charge distribution which extends in two dimensions, we wish to find the E field due to a large, assumed infinite, sheet of charge. The infinite extent applies to only two of a possible three spatial dimensions. We talk of the sheet being infinitely wide and infinitely long, but it may be of finite or infinite thickness. The character of the electric field depends not only on the arrangement of the charges on the sheet but on whether the sheet is composed of conducting or non-conducting material as we will see.Before beginning this examination of planar symmetry though, it's important to understand some implications of Gauss' Law (or more importantly of the symmetries implicit in the use of Gauss' Law) for charges on conductors. First, note that conducting material has the property of allowing charges (electrons are almost always the charges we are talking about for real materials, but we must keep in mind that real materials are not perfect conductors like those we will be discussing) to move freely, i.e. without resistance to their motion, through them.

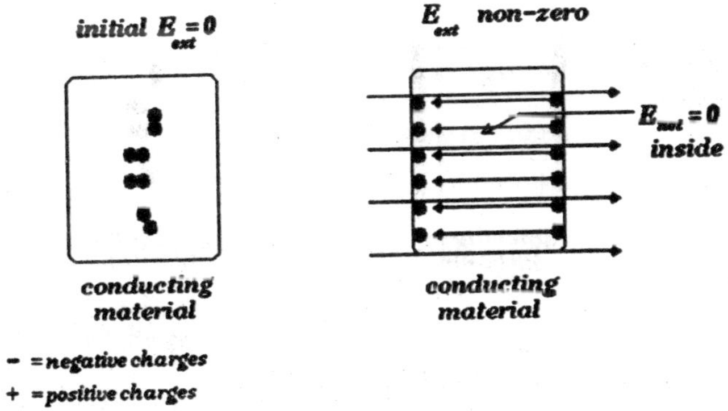

Fig. Charges Separating within a Perfectly Conducting Material.

The mobility of the charges means that they respond to any electric field in their vicinity by moving according to F = qE. If we think about the consequences of this, you'll quickly come to the conclusion that the electric field inside the bulk of any conducting material must quickly go to and stay at zero. The reasoning is simple: if charges move in response to electric fields, then the charges will only stop moving once the electric field goes to zero. However, since the negative and positive charges move in opposite directions they will themselves create an electric field. A simple picture suffices to show that the electric field created always opposes the external electric field creating the charge separation.

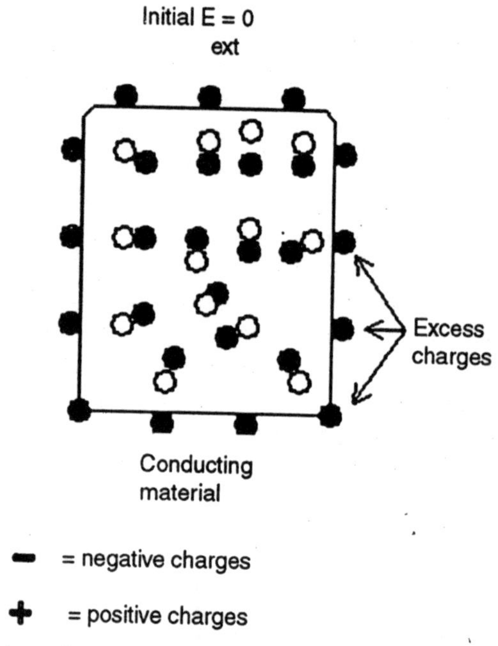

Fig. A perfect conductor with a non-zero electric charge.

The charges keep separating until they reach the surface of the material. More and more charges accumulate at the surface and create a net electric field that exactly cancels that of the external field. We assume that there are enough charges in the material so that any electric field can be neutralized. Thus, the initial electric field in the material is zero since the positive and negative charges are "combined" so as to leave no net charge to create an electric field. Once an external field is turned on, the net electric field inside the bulk material is zero since the charges separate to produce an electric field that opposes that of the external field. Later, we will see how to modify the above picture for "perfect" conductors so that we can determine what happens for real conductors.

What we have shown above works fine for conductors that have no net charge. We can also apply similar reasoning to the case of a perfect conductor with a non-zero net charge. In the absence of any external electric field, the net charge on the conductor winds up on its *outside* surface. This maximizes the separation between the charges and minimizes the forces acting on them.

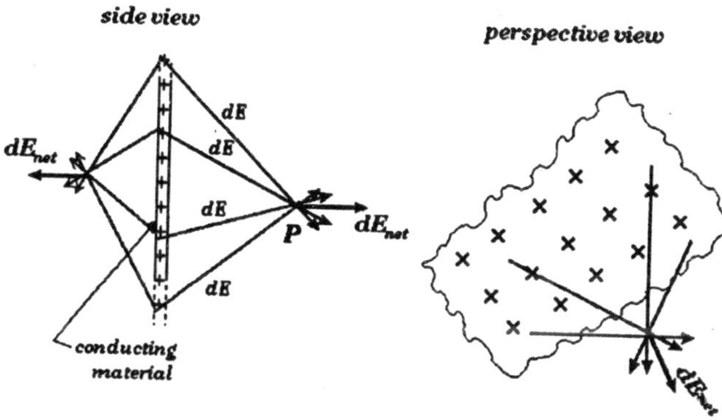

Fig. The electric field near an infinite sheet of charge.

Note that the charges inside the conductor don't separate, but instead orient themselves so as to still generate a net field that opposes that due to the external charges, thereby neutralizing them so that net electric field inside the bulk material remains zero. Again, it is assumed that there are enough charges in the conducting bulk to neutralize the field due to any amount of external charge placed on the conducting material. Again, we will talk more of this later, but for real materials we know that the number of internal charges is not infinite. So, the electric field can penetrate a certain very small distance into the conducting material. We refer to this as the skin depth of the material. We can assume the skin depth to be of zero width. This is an acceptable approximation for many "good" conductors.

To return to our original discussion of planar symmetry, we go first to the picture of what happens to the net field produced by charges confined to a sheet of infinitesimal thickness. The figure shows the result of adding up the E fields from charges at various points on the sheet. As in the spherical and cylindrical cases, we see a symmetry that allows us to cancel components of the net E field that are *parallel* to the sheet and only keep those that are perpendicular to the sheet.

Before going through this problem with Gauss' Law, we can take the "old" approach of Coulomb's Law. At least we can now make use of a previous result to reduce our workload. Remember that we have already calculated the magnitude and direction of the E field due to a wire. We now make use of this result. Consider the sheet as being composed of an infinite number of infinitesimally thin wires, each with width dx. We wish to find the electric field at a point P located a distance b from the sheet.

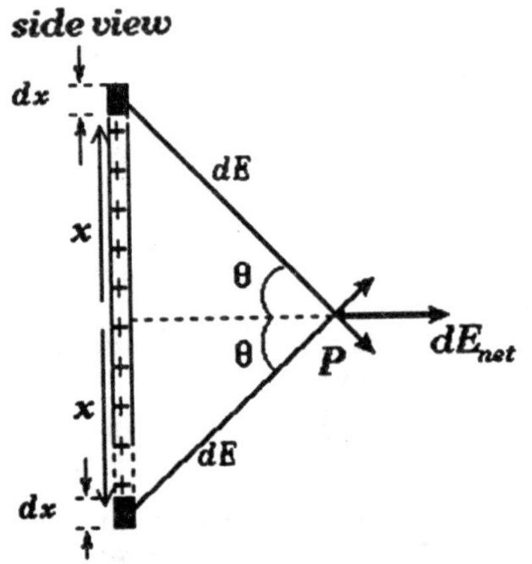

Fig. Calculating the E field near an infinite sheet of charge.

As usual, we note that the symmetry, planar symmetry in this case, allows us to ignore components of the net E field as these cancel. The components perpendicular to the sheet add, so, since the sheet is infinite in extent width and length, the E field is perpendicular to the sheet no matter what point

P we pick. Let's choose, as a reference point, the spot on the charged sheet which is closest to the point P where we want to know the field magnitude (marked as point O in above figure. This point lies on a line through P which is perpendicular to the sheet. If we choose "wires" which are equidistant from point P, then the contributions to the field at point P from components which are parallel to the charged sheet will exactly cancel. The perpendicular components are each dE cosq where q is the angle between a line from the wire to point P and the line from point O to point PAccording to the result derived previously, the electric field at point P due to a single wire is

$$E = \frac{\sigma dx}{2\pi\varepsilon_0 b}$$

Note that we have replaced the linear charge density of the previous lecture with σdx. The distance from the wire to point P, h, has been replaced by the distance r. Now the integration over all the wires can occur. $E_{net} = \sigma / 2\varepsilon_0$, where you should note that the integral can be evaluated using Maple or by consulting a table of integrals. The above result works for an infinitesimally thin sheet of conducting material or for any finite thickness sheet of non-conducting material since the E field configuration will be the same.

If we apply Gauss' Law to the same case, we can arrive at the solution considerably faster. Imagine a Gaussian surface in the form of a block with a rectangular cross-section.

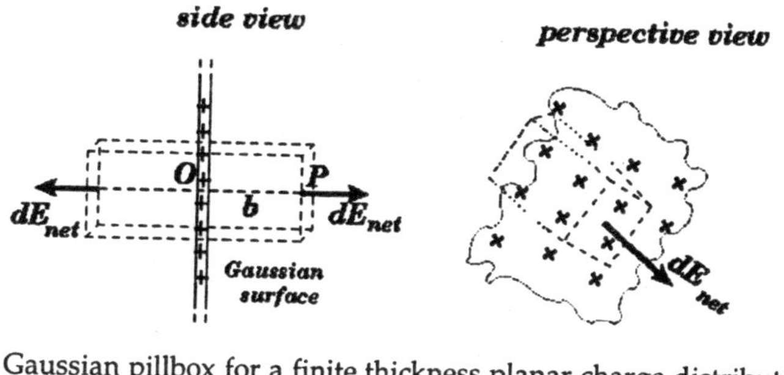

Fig. Gaussian pillbox for a finite thickness planar charge distribution

The electric field through the Gaussian surface is parallel to the sides and perpendicular to both the endcaps. The result is that the flux through the sides is zero while the flux through the endcaps is EA + EA.

with A being the cross-sectional area and where we take advantage of our previous calculation with Coulomb's Law for the infinitesimally thick conducting sheet (or finite thickness non-conducting sheet) to state that the electric field should be constant in direction and magnitude on either side of the sheet. Hence, Gauss' Law states that

$$\text{flux} = \Phi = 2EA = \frac{q_{enc}}{\varepsilon_0}$$

Since the charge enclosed by the Gaussian surface is

$$q_{enc} = \sigma \cdot AE = \frac{\sigma \cdot A}{2A\varepsilon_0} = \frac{\sigma}{2\varepsilon_0}$$

This is the same result as was derived using Coulomb's Law as we expect. The difference in effort to get this result is impressive, although we did depend on having seen the symmetry that allowed us to relate the flux to the electric field in this case.

The real payoff of using Gauss' Law comes when we apply the same thought process to the case for a conducting sheet of semi-infinite thickness.

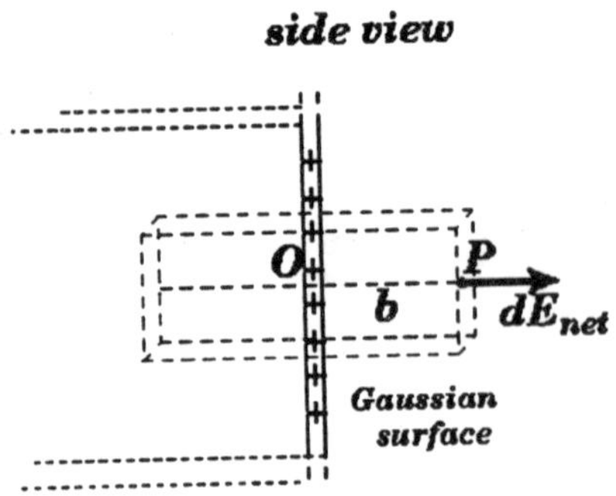

Fig. Gauss' Law applied to a slab of semi-infinite thickness.

Using Gauss' Law with the same arrangement as before, we note that now only one endcap receives a net flux since the endcap inside the conducting material sits in a zero field region. Hence the magnitude of the E field in this case is

$$\text{flux} = \Phi = EA = \frac{q_{enc}}{\varepsilon_0} = \frac{\sigma}{\varepsilon_0}$$

so the field is twice the magnitude of the non-conducting sheet of finite thickness or the conducting sheet of infinitesimal thickness.

THE FINITE THICKNESS INFINITE SHEET

Even though we've used planar symmetry to derive results for a couple of different examples of infinite sheets of charge, we should consider one more example, not for derivation, but strictly for application of the ideas learned so far. Consider the following problem from the 1991 midterm I.

An infinite slab of insulating material has a uniform volume charge density of 2.0×10^{-6} C/m^3. The surfaces of the slab are located at x = +1 cm and x = –1 cm.

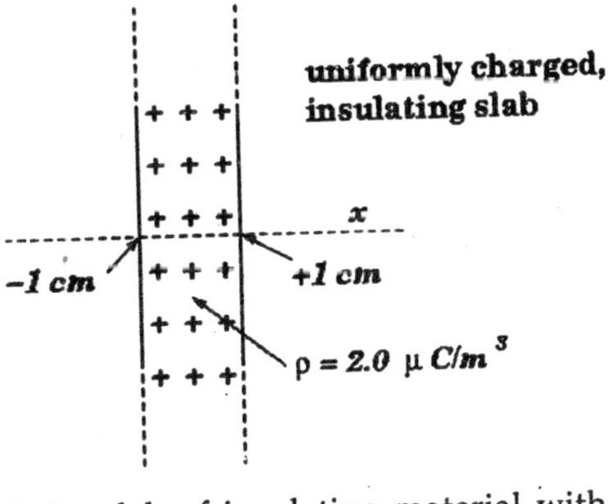

An infinite slab of insulating material with a uniform charge density throughout.

Find the electric field at:

1. $x = 2.0$ cm
2. $x = 0$

3. $x = 0.7$ cm

We need to draw a Gaussian surface which reflects the symmetry. From what we just covered in the last lecture, we know that the electric field should be constant in magnitude and direction *outside* the nonconducting material. For positive charge, the electric field is directly perpendicularly away from the surface of the material. We can use a Gaussian surface of any shape we desire, but let's choose one of rectangular cross-section as we have thus far in our derivations. That means we have a situation that looks as follows:

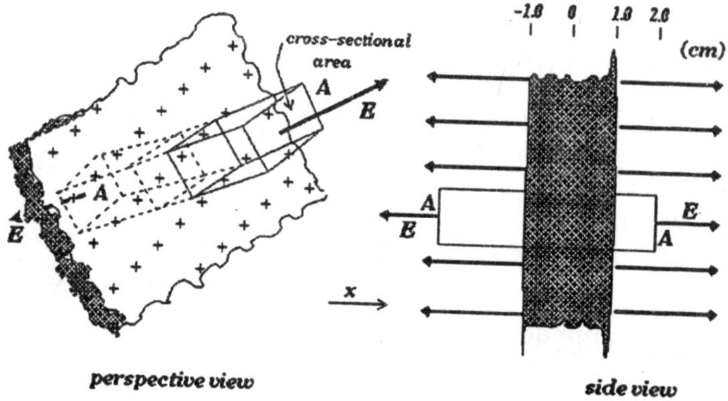

Fig. A rectangular Gaussian surface in the vicinity of a charged slab of infinite extent.

Note that the Gaussian surface on the +x side extends only to the 2.0 cm mark for which we desire the field magnitude (direction is already known to be along +x, i.e. perpendicular to the surface at +1.0 cm). On the -x side we need only note that the Gaussian surface must extend <i>outside</i> the non-conductor to preserve the symmetry of the electric field. The reason is that we know the field has constant magnitude and direction outside the nonconducting material, hence it does not matter where we place the endcap of our Gaussian surface as long as it is at x less than -1 (i.e. x = -3 cm or x = -10 cm work equally well). Since the field has the same magnitude on either side of the nonconducting material and we have drawn the surface so the sides of the Gaussian "box" are parallel to the direction of the field for either +x or –x, the net flux through the Gaussian surface is 2E·A where A is the cross-sectional area of the Gaussian box. To get the charge enclosed by the box, we use the volume charge density, which we call rho, and the volume contained by the part of the Gaussian box which is inside the nonconducting material. This total charge is $q_{enclosed} = \rho \cdot A \cdot t$ where t is the thickness of the non-conducting material. From the information given, t is obviously 2 cm. Hence, it is quick to apply Gauss' Law: 2EA = $q_{enclosed}/\varepsilon_0$, E= 2260 N/C

1. By symmetry, at the centre of the non-conducting material must have zero net electric field. Fields from each half of the slab would cancel.
2. Now we have a choice of how to draw the Gaussian surface. We can choose a symmetric arrangement in which the two endcaps lie at -0.7 cm and +0.7 cm or one in which the two endcaps lies at 0 and +0.7 cm as shown in the figure.

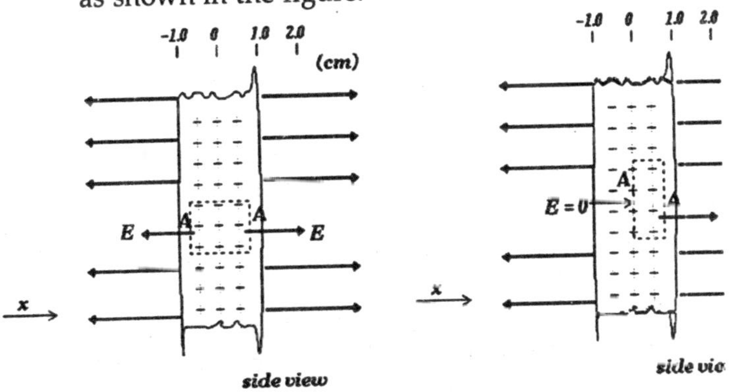

Fig. A Gaussian surface partially embedded in a charged slab of infinite extent.

The reason either of these works is because we can evaluate the flux and separate out the contribution of E to the flux in either case. For the first case, the net flux through the surface is just 2EA since, by symmetry, the E field will be the same magnitude at the same distance from the symmetry point of x = 0. In the second case, the flux is just through the right endcap since the magnitude of E at x = 0 (location of the left endcap in this case) is zero. Hence the flux would be EA. Now note that we get the same result for either case. For example, in the first case $2EA = q_{enclosed}/\varepsilon_0$, E= 1582 N/C

We can easily verify that the second case gives exactly the same result. Remember that the flux is one-half of what it is in the case just shown, but that the volume enclosed and therefore charge enclosed in the second case is also half of what we get for the first case.

ELECTRIC POTENTIAL

We are still operating under the assumption that the electric field provides a convenient "mechanism" for dealing with electrical phenomena. Even though electrical forces are what we observe, the electric field is assumed to the be the "agent" by which the force is manifested on charged particles. However, just as we found with Newton's Laws, it becomes increasingly clear, with some experience, that we would prefer, for practical reasons, to adopt a way of describing the electric field in terms of scalar quantities rather than through the vector nature of the E field itself. With Newton's Laws of Motion, it turned out to be very useful to define the concepts of *work* and *kinetic energy*. These definitions, namely $W = F \cdot ds, K = 1/2mv^2$ relate work to an integral over a path (which is broken into infinitesimal increments of displacement, ds) and the kinetic energy to the mass and velocity of a point particle. We note that these two are related according to the *work-energy theorem*, $W = K_f - K_0$. Furthermore, for work done by conservative forces, we can define a potential energy, $U(x) = F(x)\,dx$

The generalized work-energy theorem relates changes in potential energy to work and hence to changes in kinetic energy for the case of conservative forces

$$W = -\Delta U = \Delta K \; K_0 + U_0 = K_f + U_f$$

The potential energy and kinetic energy allowed us to work with scalar quantities dealing with motion and the change in motions caused by forces. We can develop an analogous scalar quantity to the electric field. It is called the electric potential and is usually represented as V (due, as we will see, to the common name for the unit of electric potential, the volt). The relationships we seek then look, schematically, as follows

$$\begin{array}{ccc} \boldsymbol{E} & \longleftrightarrow & \boldsymbol{F} \\ \updownarrow & & \updownarrow \\ \boldsymbol{V} & \longleftrightarrow & \boldsymbol{U} \end{array}$$

Fig. Relationships between scalar and vector quantities related to static electric fields.

There's nothing especially tricky here. V becomes the scalar quantity of choice to use for problems involving the electric field. The idea is that the scalar nature of V will be generally easier, and therefore more practical, for uses than E just as potential energy is more useful for many problems than force, F. However, for V to be useful, we need a definition for it which allows us to consistently realise the pictorial relationship between force, potential energy, electric field, and electric potential that is drawn above.

The definition which works is

$$U(x) = –F(x)\ dx = –q_0E_x\ dx$$

$$\Delta V – W/q_0 = \Delta U/q_0\ V = U/q_0$$

That is, we state that the work per unit charge is the change in potential. This is entirely consistent with thinking about V in the same way we think about height for the gravitational field. The higher the value of V with respect to some fixed reference, the more potential energy (for a positive charge) there is in the electric field, just as increasing the height of a mass raises the potential energy in the gravitational field. One consequence of this definition is the relationship between electric potential and electric field. From the generalized work-energy theorem and the definition of electric potential, we have

$$= –W/q_0$$

$$= \Delta U/q_0$$

$$= -1/q_0 \text{ F.ds}$$
$$= -1/q_0 \text{ q E.ds}$$
$$\Delta V = -\text{E.ds}$$

The s refers to a path along a electric field line. V becomes the scalar quantity of choice to use for problems involving the electric field. The idea is that the scalar nature of V will be generally easier, and therefore more practical, for uses than E just as potential energy is more useful for many problems than force, F.

The Electric Potential

We will find many uses for the electric potential in the near future. For now, we note that it is indeed easier to work with than the electric field because of its scalar nature. The first way to calculate the potential is through the use of integral as defined above. We can apply this to a case for which we know the electric field, namely a finite length charged, rod as shown below. We wish to find the electric potential at a point P which is a distance h above the midpoint of the rod of length L.

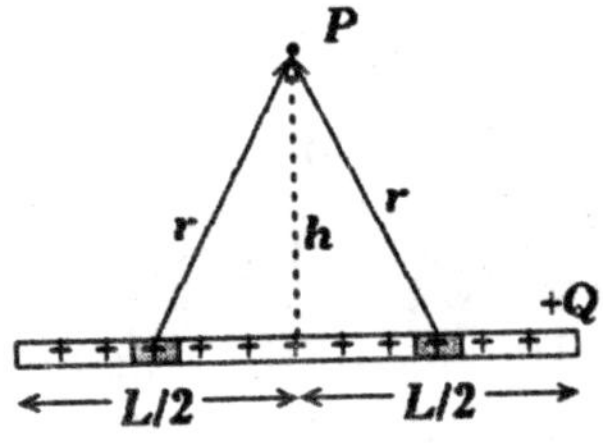

Fig. Finding the electric potential over the midpoint of a finite length, charged rod.

If we wish to find the potential at point P relative to infinity, then we should integrate the dot product of E with ds over a path from infinity in to point P. The easiest path to integrate is a straight-line path as the E field is always vertically upward and opposite to a straight line path from infinity down to point P. So, noting that y is the integration variable and goes from the line charge *out* to infinity (and therefore is opposite the ds direction). $V_P = -E \cdot ds$

We can also consider doing this calculation another way. If we make use of the techniques of calculus, we can consider dividing the rod into infinitesimal lengths of charge, dq, then considering the potential at point P for each. Adding up the potential for all the dq's by integration should yield the same result as above. Since the infinitesimal lengths can be considered as point charges, we need to know the potential at point P, relative to infinity, for a point charge dq. First, consider any point charge, Q, in space. Then the electric potential at a point a distance r away can be calculated using y as the integration variable as follows. The electric field due to a point charge is radial, so we can integrate over a straight line path along a radius from infinity to r. with x being the distance of dq from the midpoint of the line segment charge. Even though we are not dealing with vector components, we can still take advantage of symmetry in that we see that the right and left halfs of the line segment (i.e. to the right or left of the midpoint of the line segment) are completely symmetrical: they give equal contributions to the potential. As expected, this gives the same result. This example brings up something of a lesson here. If you wish to find the potential of a charge distribution of finite size, you can always break the distribution into infinitesimal pieces, treat each as a point charge for getting the potential at the desired point, then integrate over the whole distribution to get the net potential. If you know the electric field distribution of a charge distribution or can easily calculate it, you can find the potential relative to some reference point by integration of the electric field over some path. The fact that you can use any path means you always want to choose the path that minimizes the difficulty of the integral.

Electric Potential for a Finite Thickness, Infinite Sheet

We now want to apply what we developed at the end of the last lecture in deciding which of our two methods to use for calculating the electric potential due to a charge distribution. Let's consider the case of the infinite, charged, non-conducting sheet.

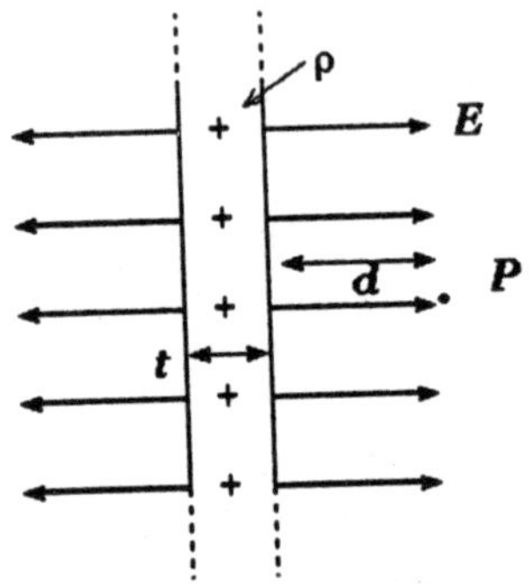

Fig. An infinite plane with volume charge density r and finite thickness t.

If we assume the sheet has finite thickness, t, and a volume charge density, r, then we know,that the electric field has a magnitude $E = \rho t/2\varepsilon_0$.

At any point outside the sheet. The electric field is directed perpendicularly away from the sheet (for a positively charged sheet). So, if we want to know the potential at point P, *relative to the right surface of the sheet,* then we can evaluate the potential via a straight line path from the right face of the sheet out to point P. $V_P = -Ed$

If we considered trying to break the infinite sheet into infinitesimal charges and integrating, we would get the same answer, but the effort we would need to expend is significantly greater. Clearly this problem is best done using our previous knowledge of the properties of the electric field.

Potential of a Charged Ring

Consider a thin ring of plastic with a radius R and a uniformly distributed net charge –Q as shown in the figure.

We can quickly surmise the answer to the first part of the problem. By symmetry, the electric field at point P must be zero. Since we can break the circle into infinitesimal pieces, dQ, with arc-length ds, we can quickly see that every piece has another piece, diametrically opposite it, which delivers a contribution to the net field which has equal magnitude and opposite direction.

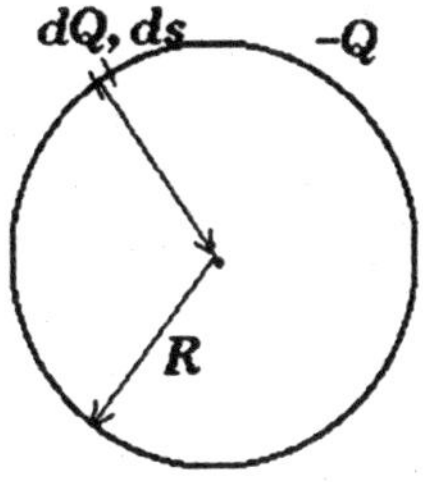

Fig. A thin ring of radius R and uniformly distributed charge -Q.

We would be unwise to be so hasty with the second part of this question however. The electric potential, after all, is evaluated by integrating over a path from infinity to point P. Even though the electric field at point P is zero, it is definitely not zero for any arbitrary position which is not at point P. Hence we have no reason to expect that we would get zero when integrating E · ds, and, in fact, we don't. In this case calculating using the E field is tedious compared to just integrating over the contributions due to each infinitesimal piece of the ring, dQ. To derive dQ, use the linear charge density method, l = Q/(2πR), where 2πR is the circumference, or total length, of the circle. So, dQ = λds. V_P = –kQ/R,The answer is distinctly not zero. The lesson is, be careful! Knowing E at a point says nothing about the value of V at that point. Although we can certainly say that, in regions in which E = 0, the potential does not change.

Electric Potential Inside a Sphcrical Thin Shell Conductor

Consider a hollow metal sphere of radius R which has a net charge +Q. Find the electric potential, relative to infinity, at the centre of the sphere.

We know that a conductor must have all of its net charge sitting on a surface. In this case, its the outer surface of the sphere. Furthermore, we know the electric field is the same as for a point charge sitting at the centre of the sphere for points which are *outside* the sphere.

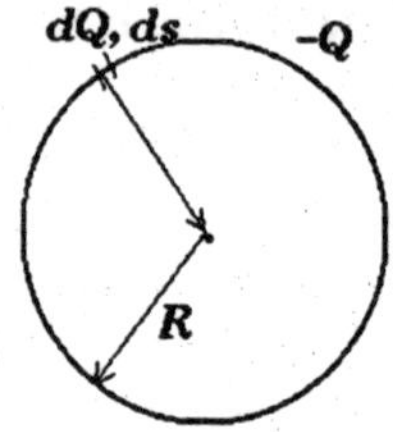

Fig. A thin spherical shell of radius R and charge +Q.

Inside the sphere, the electric field is zero since any Gaussian surface we draw which is completely contained inside the sphere would contain no net charge. If we set the origin of our coordinate system as being at the sphere centre, then we need to integrate the electric field over a path (it might as well be straight-line since that's the easiest to integrate) from infinity to the sphere and from the sphere surface to 0 (since the electric field is different inside than outside). So, remembering that the path is opposite to the direction of the E field (radially inward for the path, radially outward for E). $V_P = -kQ/R$.

Electric Potential Inside a Hollow Metal Sphere

Suppose we allowed the shell from the previous problem to have a finite thickness? How does that change our result for the electric potential inside the shell? Assume that we have a spherical shell with outer radius b and inner radius a as shown in figure . What is the electric potential, relative to zero at infinity, fcr a point P at radius r < a?

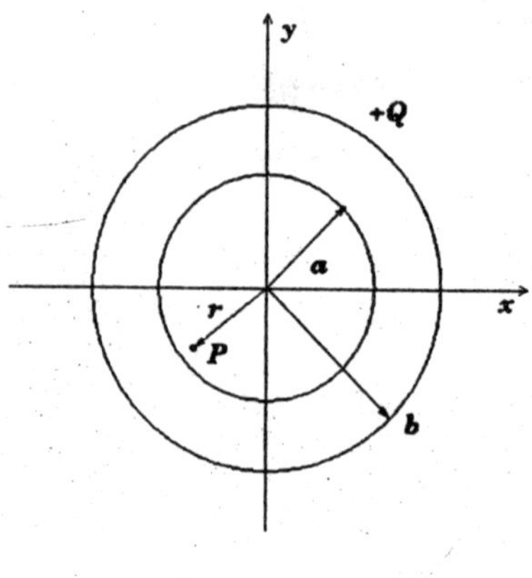

A spherical, conducting shell with inner radius a and outer radius b. We wish to find the electric potential for a point located at r < a.

The approach to the solution is identical to our previous example for r > b. That is to say, the potential of the outer radius of the conducting shell is

$$V_b = \int \vec{E} \cdot d\vec{s} = -\int_{\infty}^{b} \frac{kQ}{r^2} dr = \frac{kQ}{b}$$

How about the region a < r < b? The electric field in the bulk of a conductor must be zero, so the electric potential cannot change for this region. Hence,

$$V_{a<r<b} = V_b = \frac{kQ}{b}$$

Finally, for the region r < a we also have no electric field, hence, $V_{r<a} = V_{a<r<b} = V_b = kQ/b$,Therefore, for any point inside the shell (or within the conducting material of the shell), the potential is the same as the outside surface. This includes the potential at the centre of the shell, hence the result is unchanged!

Equipotential

We take from the previous example a potent lesson: any path for which conducting material provides a continuous link must be at the *same* electric potential. We can make use of this to work out problems that might seem intractable at first. For example, consider two conducting spheres with differing radii R_a and R_b sitting on insulating stands far apart. The sphere with radius R_a has an electric charge +Q. If we connect a thin, conducting line between the spheres, then disconnect it, what are the charges on the two spheres?

The conducting line brings the two spheres to the same electric potential. Since the sphere with radius R_b was initially uncharged, bringing the two spheres to the same electric potential requires some of the charge on the R_a sphere to go to the R_b sphere. This reduces the potential on R_a and increases it on R_b until the two come to the same electric potential relative to infinity. We can calculate what this charge is by noting that the final charges must satisfy charge conservation such that $Q_a + Q_b = Q \Rightarrow Q_a = Q - Q_b$.

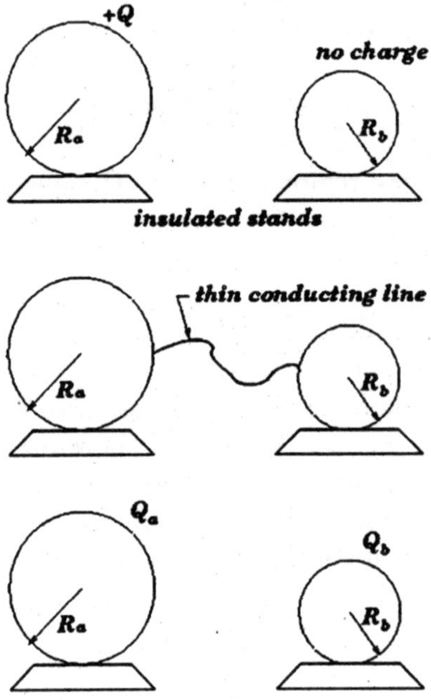

Fig. Two shells on insulated stands are briefly connected by a conducting line, then disconnected.

The final condition of the potentials being equal is easily represented provided we have the spheres separated by a considerable distance. This ensures that the electric field of one sphere does not have much influence on the charge distribution of the other sphere. For the potentials to be equal, we must have

CAPACITANCE

Methods of storing energy for the purpose of doing work and directly studying electric fields and their effect on matter was an important focus of research in the mid-1700's. Methods of storing electrical charges for long periods of time were developed, the most important being the creation of the Leydan jar. Methods of improving the Leydan jar by simplifying its construction and increasing its *capacity* for storing charge soon resulted in devices consisting of just two flat metal sheets separated by air. One of the sheets can be charged, then disconnected from all electrically conducting material. The charged sheet induces an equal charge on the second sheet. In its simplest model, any two oppositely charged metal objects separated by a distance forms a "condenser" of charge or a capacitor, as we now say. In order to provide a "figure of merit" for how effective any type of capacitor is at holding charge, we define the capacitance through the relationship $C = Q/V$

The higher the capacitance, the more effective the capacitor is in the following sense: if we look at the relationship between C, Q, and V as defined, we note that the more charge we store per unit of V, the higher the value of C. Since V relates to the amount of work needed to produce the charge Q on the plates of the capacitor, having a high capacitance means we store more charge for less work done in charging. As we will show, capacitance also is a measure of how effective the capacitor is at storing energy, and hence how effective the capacitor can be at *doing* work after it is charged.

First, let's consider the parallel-plate capacitor as shown in figure.

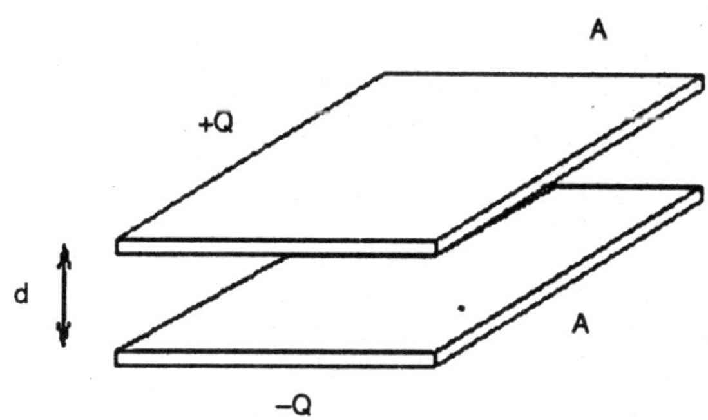

Fig. Two thin plate conductors with fixed distance d, plate area A, and charges +Q on one plate and –Q on the other form a parallel-plate capacitor.

The two conducting plates each have an area A and are separated by a distance d. By using a source of EMF or ElectroMotive Force (a device which moves charges from one point to another), we can drive electrons onto the bottom plate while removing them from the top plate. After the plates are charged, we can, if we wish, remove the EMF source and the plates, if isolated from any other conducting material will hold their charges for a considerable period of time; years in some cases.

We can evaluate the capacitance for the parallel-plate case by noting that the electric field between the plates is, according to Gauss's Law as applied in previous lectures, constant in magnitude and direction.

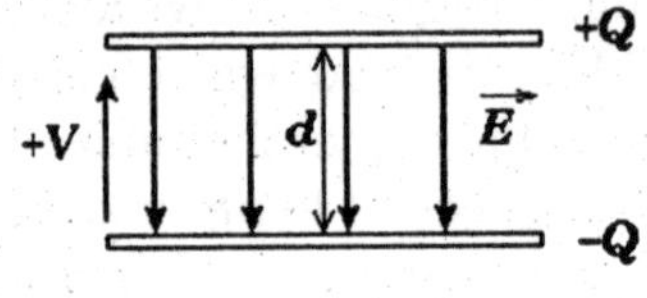

Fig. A parallel-plate capacitor with plate area A and charge Q.

Hence, we know that $E = Q/e_0A$. The voltage between the plates increases in going from the negative to the positive plate and has a magnitude V = Ed. Therefore,

$$C = Q/V = \frac{Q}{Qd/\varepsilon_0 A} = \frac{\varepsilon_0 A}{d}$$

Therefore, for the parallel-plate capacitor, the capacitance is proportional only to geometric factors (the area of the plates and the separation distance between them) and the natural constant, ε_0.

Batteries and Capacitor

We note that we are more familiar with the concept of storing electrical energy in *batteries*. Hence, it makes some sense to distinguish between batteries and capacitors. The purpose of a battery is to use chemical (or other) processes to produce the EMF needed to move charges around. This is done by using the chemical or other process to maintain a constant potential between the *poles* (i.e. the positive and negative parts of the battery). This constant potential provides the EMF for moving charges from one conductor of a capacitor to another, for example. Hence, this is the principal distinguishing feature of a battery and a capacitor: a battery maintains a *constant* potential difference between it's poles whereas a capacitor can maintain a potential difference between its positive and negatively charged conductors, but there is no need for that potential difference to be constant - the potential difference depends on the magnitude of the charge on the conductors! Thus, while batteries store energy, capacitors store both energy *and* charge. The motion of charges can be quite rapid and readily controlled hence capacitors can be used for electrical energy storage that requires timing as part of its function. Batteries are generally not good for this purpose.

Energy Storage in Capacitors

We can now talk about the energy stored in the electric field of or, equivalently, stored in the arrangement of electric charges within the capacitor. Again, we use the parallel-plate capacitor because it is reasonably simple. First, note that we will use *schematic* representations of circuit elements to save drawing time. For a source of EMF, or voltage that can move charges, we have the symbol in figure .

Fig. Symbol for a source of EMF within a circuit.

To calculate the energy stored, we hook the EMF source to a capacitor, e.g. by closing a switch that allows conducting lines to link the EMF source and a capacitor together.

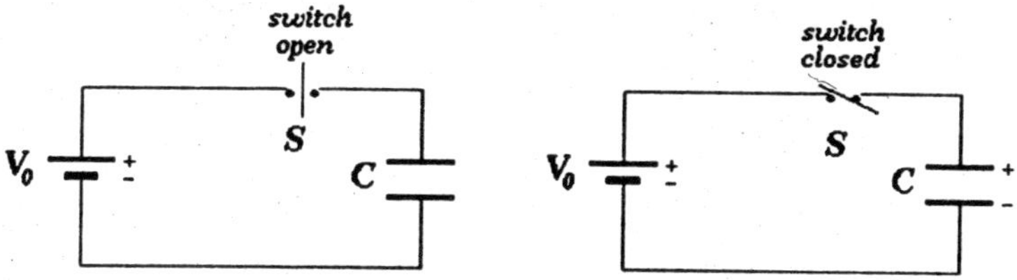

Fig. Closing a switch to charge a capacitor.

As soon as the EMF source is able to push charge onto the capacitor plates, it does so by stealing electrons from the top plate of the capacitor and pushing them onto the bottom plate. As soon as the first infinitesimal bit of charge hits the capacitor, an electric field is created between the plates. Although the field is also infinitesimally small, it still produces a voltage (again, infinitesimally small). This voltage operates *opposite* to the direction that charge is flowing, i.e. the capacitor voltage tends to push positive charge in the counter-clockwise direction while the EMF source pushes it in the clockwise direction. Since the EMF source has a much higher voltage, it succeeds in pushing more charge onto the plates of the capacitor. As the charge continues to build though, the electric field strength, and hence the voltage across the plates of the capacitor, continues to build. When the capacitor voltage equals that of the EMF source, charge flow stops and the situation is static.

At this point, the voltage across C is just V_0, the same as that across the EMF source. To find the amount of work done by the battery to bring this situation about, let's translate the above description into calc-speak, i.e say it mathematically. The first bit of charge to arrives takes no work on the part of the EMF source assuming that C is initially uncharged. This bit of charge on C causes a potential V to appear across C. The next infinitesimal bit of charge, dq, coming through the EMF source must have work done on it to push it against the potential from C. This work is just dW = dq V. Now we can integrate to see what the total work is to get V up to V_0.

$$W = ½ CV_0^2$$

We can write this in a couple of other forms if we note that $C = Q/V_0$ for this case:

$$W = \frac{1}{2}CV_0^2 = \frac{1}{2}QV_0 = \frac{1}{2}Q^2/C$$

These results were derived for the parallel-plate case but are general for a capacitor with any geometry. Since the work done is also equal to the potential energy stored for a system subject only to conservative forces, we note that the work done on the charges to get them to the capacitor plates also represents the potential energy of the charged capacitor, i.e. its ability to do work. Another task of importance will be understanding the connection between the potential energy in the arrangement of charges on the capacitor plates and its connection to the electric field. Again, we can use the parallel-plate to provide a solution which holds generally. Assume we were interested not in knowing the potential energy of C, but of the energy density or potential energy per unit volume between its plates. The potential energy divided by the volume provides this, so $u = \frac{1}{2} \varepsilon_0 E^2$

This represents the potential energy per unit volume of free space (i.e. vacuum) when an electric field of magnitude E is created in that unit volume. It also represents the work done per unit volume to establish a field of magnitude E in that space. Hence we see the role of ε_0 as signifying the effort needed to establish an electric field in a volume of empty space. Hence the name for ε_0, the permittivity of free space.

DIELECTRICS

We originally started the discussion of capacitors by noting their importance as a means for storing charge (and hence electrical energy) that could be used to do work or make studies of the electric field. As such, there was a continual push to develop capacitors of larger charge-storing capability. There are practical limitations to storing charge in that the electric field between the parallel plates of a capacitor, and hence the electric potential, goes up as the charge goes up. At a certain point, the air gap between the plates breaks down, i.e. the nitrogen in the air is ionized, becomes conducting, and therefore provides a path that enables the plates to short out. One can also decrease the distance between the plates. Since the potential across the plates goes as $Ed = Qd/\varepsilon_0 A$, reducing d decreases V for a given value of Q on the plates. This has practical limitations in keeping the distance between the plates uniform over the area. To see why keeping the plates apart is a problem, let's calculate the force between the plates. First, look at figure below showing a parallel-plate capacitor with a separation distance x between the plates. To pull the plates apart by an additional infinitesimal distance dx, we need to exert a force because the positive and negative charge plates attract one another.

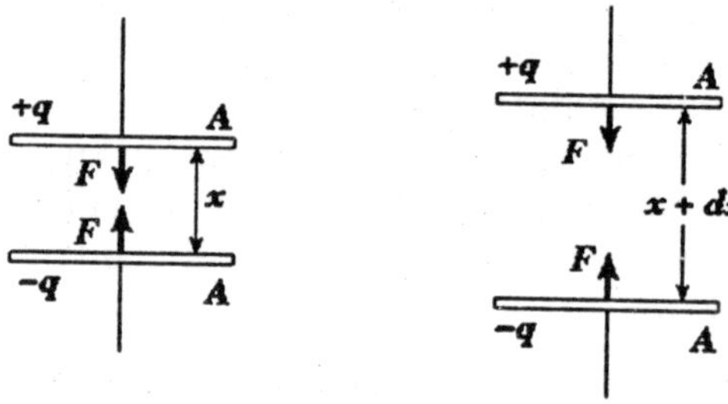

Fig. The mutual forces on capacitor plates means that it takes work to separate them.

To be formal about this, note that the potential energy of the original capacitor is

$$U = \frac{1}{2}qV = \frac{1}{2}q\frac{qx}{\varepsilon_0 A} = \frac{1}{2}\frac{q^2 x}{\varepsilon_0 A}$$

After the separation, the new potential energy is (note that the charge on the plates cannot change since there is no where for the charges to go)

$$U + dU = \frac{1}{2}qV = \frac{1}{2}q\frac{q(x+dx)}{\varepsilon_0 A} = \frac{1}{2}\frac{q^2(x+dx)}{\varepsilon_0 A}$$

The negative sign on the right-hand side indicates that the force is indeed attractive as we thought. The fact that it is non-zero indicates that plates have to be held apart, especially if the charge on them is to be large. Note also that the force increases as the square of the potential difference between the plates and inversely as the distance between them, hence attempting to increase the capacitance by reducing the distance between the plates only increases the force needed to keep them apart.

So, how do we get around the essential constraints on capacitance? First, note what happens if we take any single, charged parallel-plate capacitor and insert a slab of conducting material into the space between the plates, but do it such that the conducting material does not touch either plate.

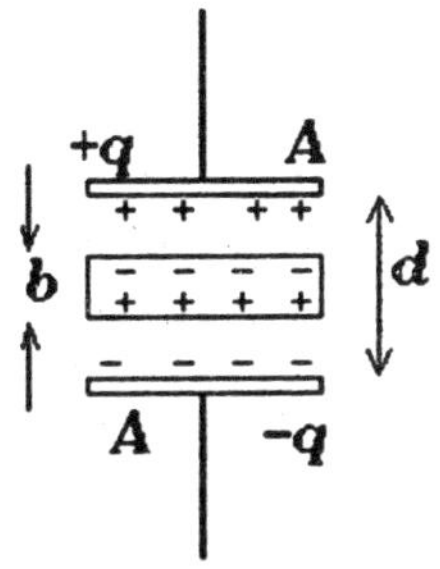

Conducting material placed inside a capacitor effectively reduces the distance between conductors and thereby increases capacitance for a given charge on the conducting plates.

The conducting material assumes the charge separation as shown in above figure. The electric field above and below the conductor remains the same as before the conducting material was introduced, $E = q/\varepsilon_0 A$ because the electric field in the vicinity of the charged plates is unaffected by the presence of the conducting material. The electric field in the conducting material is zero, so the potential from the bottom to the top of the conducting material is the same. Therefore, the potential difference between the bottom and top plates are (note that the spaces between the conductor and the plates are both (d – b)/2)

$$V = \frac{q(d-b)}{2\varepsilon_0 A} + \frac{q(d-b)}{2\varepsilon_0 A} = \frac{d(d-b)}{\varepsilon_0 A}$$

If we find the capacitance for the arrangement with the conductor inside, we get

$$C = \frac{q}{V} = \frac{\varepsilon_0 A}{d-b}$$

If we compare this to the original capacitance without the conductor in the middle, $C = q/V = \varepsilon_0 A/d$, we see that the capacitance has gone up with the introduction of the conductor. Effectively, we have reduced the gap between conducting surfaces.

This still doesn't help, technologically speaking, with making better capacitances, because we still have to keep the conductor from touching the plates of the capacitor. To get around this, we need something that reduces the electric field magnitude in the volume of space between the plates without allowing for conduction between the plates. The solution is to fill the space between the plates with a dielectric material. Dielectrics are non-conductors that are formulated from molecules which either have permanent electric dipole moments or can be induced to have dipole moments in the presence of an external electric field.

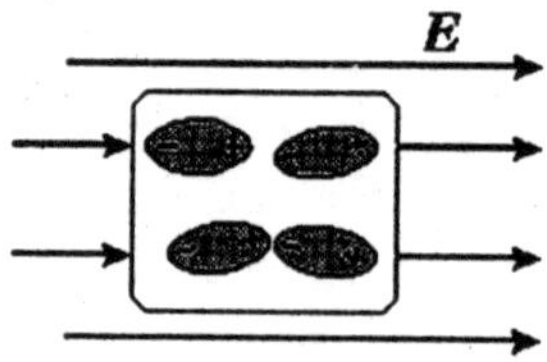

Fig. A dielectric in an external electric field.

These dielectric molecules align partially with any external electric field. They generate a field of their own within the material. This field opposes that of the external field and therefore reduces the magnitude of E within the material. While not quite as effective as a conductor, this material does not conduct charge between the plates of a conductor and still reduces the voltage as a function of charge on the plates. The reduction factor is referred to as the dielectric constant of the material and is usually represented by the letter K. The electric field produced by the charge on the plates of the capacitor remains $E_0 = q/\varepsilon_0 A$ in any region between the plates which is not occupied by dielectric. In the region which is occupied by dielectric material, the electric field is E} = E_0/K with K > 1, i.e. the electric field magnitude is reduced. Since the potential difference between the plates goes as the electric field times the distance, the lower E} (and therefore the higher K is), the more charge can be stored on the plates for a given potential difference. One way to think about the proper use of K is to note that any formula derived for capacitance can be adopted to the case in which a dielectric completely fills the space between the plates of the capacitor by simply replacing every instance of the permittivity, e_0, with the value Ke_0.

Hence, the voltage between the plates of a parallel-plate capacitor in which a material with dielectric constant K fills the volume between the plates with separation d is

$$V = \frac{qd}{K\varepsilon_0 A}$$

The capacitance is $C = K\varepsilon_0 A/d$, etc. If we apply the above consideration of calculating quantities by replacing e_0 with Ke_0 to the case in which an empty capacitor with charge q on its plates has a dielectric introduced into it, we find an anomaly.

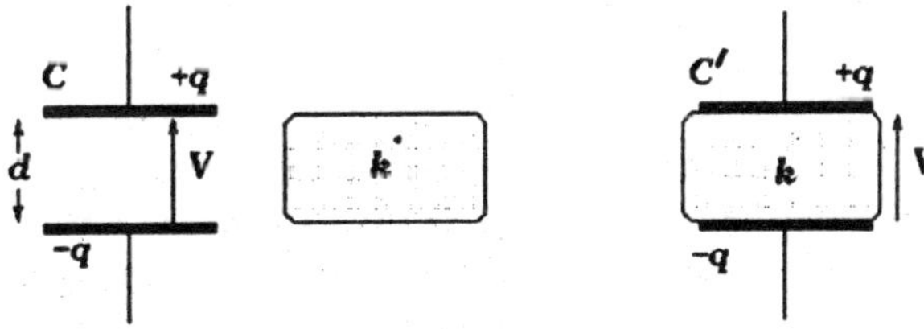

A dielectric material is used to fill the space between parallel plates of a capacitor. The capacitance and voltage are modified but the electric charge on the conducting plates themselves is unchanged. The net surface charge density is reduced due to the dielectric.

Before and after the dielectric is inserted, we have, if we let $\varepsilon = K\varepsilon_0$,

Before insertion	After insertion
$C = (\varepsilon_0 A)/d$	$C\} = (\varepsilon A)/d$
$V = qd/(\varepsilon_0 A)$	$V\} = qd/(\varepsilon A)$
$U = {}^1/_2 CV^2 = (q^2 d)/(\varepsilon_0 A)$	$U\} = (q^2 d)/(\varepsilon A)$

Notice that the energy stored has decreased by a factor of K! Where did this energy go? To understand, we again have to turn to forces exerted by the capacitor plates. For any real capacitor we must have fields which bow out at the edges, the so-called fringe fields.

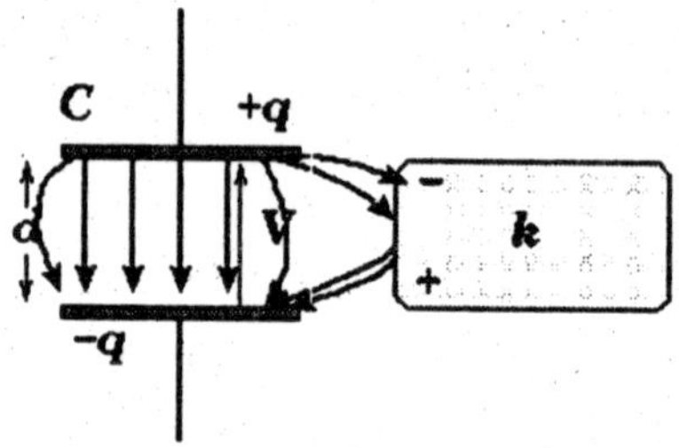

The fringe fields at the edge of the capacitor attract the dielectric material and do work to pull the dielectric into the space between the capacitor plates.

As the dielectric is introduced into the fringe field, the dielectric molecules start to align with the external field. This creates an attractive force as the plus charges of the dielectrics are aligned so that they are closest to the negative plate of the capacitor (and of course the same process occurs for the negative dielectric charges and positive plate of the capacitor). The horizontal components of the fringe field yield a horizontal, attractive force which pulls the dielectric into the space between the capacitor plates. The work done by this force in accelerating the dielectric into the capacitor volume decreases the energy stored in the capacitor, thereby lowering the energy stored in the capacitor. Some energy is also converted into electromagnetic radiation and into heat.

COMBINATIONS OF CAPACITORS

Parallel Combinations

Having determined the basics for determing capacitance for various capacitor geometries, it is now time to consider cases in which more than one capacitor may be involved in a circuit. We'll restrict the derivations to examples using parallel-plate capacitors for mathematical ease although the results will be quite general.Let's start by looking at the parallel combination. In this case, we have capacitors which have the same potential difference across their plates.

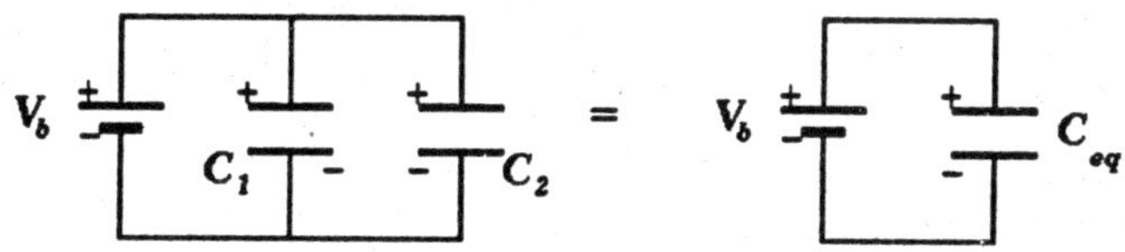

Fig. Two capacitors in parallel and a circuit with an equivalent capacitance.

For such combinations, we can simplify matters a bit by finding an *equivalent* capacitor whose characteristics make exactly the same demands from the EMF source in charging up the capacitors (presuming they had the same initial charge before the EMF source was connected). For this to be the case, the EMF source will have to push the same amount of charge onto the equivalent capacitor as it does for the two original capacitors it replaces. The work done by the emf source should be the same as well. This means that the potential difference created on the equivalent capacitor should be the same as for the two capacitors it replaces. We can now ask what the capacitance of the C_{eq} must be for these conditions to hold. Since the potentials across C_1, C_2, and C_{eq} must all be the same and the charge on the equivalent capacitor must be the same as on the two individual capacitors in the original circuit, we have $Q_{eq} = Q_1 + Q_2$, $C_{eq}V_b = C_1V_1 + C_2V_2 = C_1V_b + C_2V_b$, $C_{eq} = C_1 + C_2$

So the equivalent capacitor needs to have a capacitance which is the sum of the two capacitors it replaces. This result can easily be extended to any number of capacitors so that the result of n capacitors connected in parallel is $C_{eq} = C_1 + C_2 + C_3 + ... + C_n$

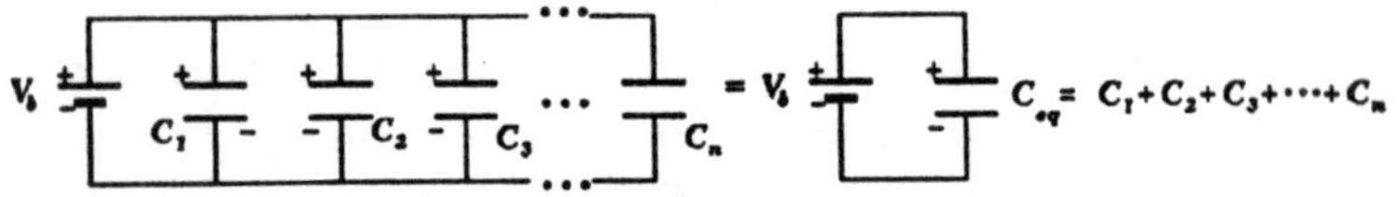

Fig. Extending the parallel capacitor formula to any number of capacitors.

Series Combinations

We can also consider capacitors arranged so that the potential across the combination is equal to the sum of the potential difference across each as shown in figure

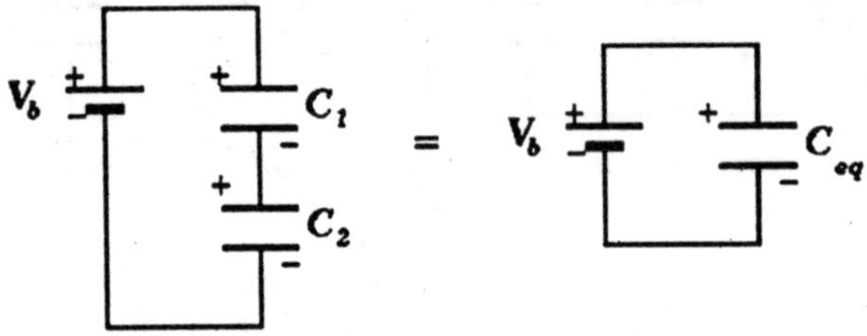

Fig. A series arrangement of capacitors and a circuit with equivalent capacitance.

As with the parallel combination, we want to find the equivalent capacitance that can replace the pair of capacitors here without changing anything that the EMF source must do to produce full charge on the combination. In this case, it is the charges that must be equal magnitude on the combination since we note that the positive pole of the EMF source contacts one plate of one capacitor in the pair while the negative pole contacts one plate of the other capacitor. In each case, the plate connected to the battery pole *induces* the equal magnitude charge on the other plate for that capacitor. Since the inner plates are electrically isolated from the world, they maintain a net charge of zero (assuming all capacitors were initially uncharged before the EMF source was connected). Thus, $q_1 = q_2 = q_{eq}$ for the combination. The potential in going from the bottom plate of C_2 to the top plate of C_1 is $V_2 + V_1$, hence $V_b = V_1 + V_2$, $q_{eq}/C_{eq} = q_1/C_1 + q_2/C_2$, $1/C_{eq} = 1/C_1 + 1/C_2$,

This can also be extended to any number of capacitors in parallel. For example, for n capacitors in series,

$$\frac{1}{C_{eq}} = \frac{1}{C_1} + \frac{1}{C_2} + \frac{1}{C_3} + \ldots + \frac{1}{C_n}$$

CURRENT AND RESISTANCE

Much of our determination of the properties of systems involving charges has been with the implicit assumption that those charges were *static*. Just as in mechanics, we can learn a lot by understanding the static conditions of objects. Note that static here did not imply that objects didn't move. The equation for motion apply to any object that has constant acceleration, i.e. the only thing that had to be static about the system was its acceleration. We could determine a number of things about the world just from this. To truly understand motion, however, we needed Newton's Laws as a way of describing dynamics. This opened up a whole new universe of understanding which went well beyond Galileo's equations for motion. Phenomena that were complete mysteries before, e.g. the orbits of planets, suddenly had precise descriptions in terms of Newton's Equation of Motion.We have the same situation with electricity. We will see later in the course that electrodynamics offers explanations of phenomena that could not even be guessed at with just the applications of Gauss' and Coulomb's Laws. To get to the point of understanding this though, we have to start from the beginning. We begin by formally defining current as the time rate of change of charge through a conductor. Thus, if we have a conductor material of any cross-section, we can define the current in that material as the amount of charge that passes through a hypothetical plane that cuts all the way through the material.

In the figure above, each of the hypothetical planes above sees the same current, i = dq/dt, go by it. We define the unit of current as the Ampere and abbreviate it as A. 1 A = 1 coulomb/ sec. Since charge is conserved, note that current, the time rate of change of charge, must also be a conserved quantity. If were not, then charges could accumulate or disappear at some point in the material. This is assumed not to happen for conducting material nor for electrical devices. Currents flow as the result of potential differences between positions.

RESISTANCE

Currents are not dependent solely on potential differences. An object capable of carrying current can be characterized by its resistance where resistance is defined as

$$R = \frac{V}{i}$$

The resistance of an object depends on its geometry and its resistivity. Resistivity is a property of the material itself and is the same no matter what the shape or dimension of the material. The symbol for resistivity is the Greek letter rho. For an object made from material with resistivity rho,

$$R = \frac{\rho L}{A}$$

where L is the length of the object and A is its cross-sectional area which is assumed to be constant throughout the object. If A it is not constant, then we need to do an integration over infinitesimally small distances of the material where we can assume that the cross-sectional area is a constant.

For this course, unless explicitly stated otherwise, you can always assume that there is a linear connection between potential difference, current, and resistance in which V = i R

Although this appears to be identical to the definition of R, keep in mind that this relationship assumes that potential difference and current are linearly related over some reasonably large set of possible V's and i'. Semiconductor materials, as an example, display remarkably (and most usefully) non-linear relationships between potential difference and current. The above equation of V = i*R is called Ohm's Law because it holds reasonably well for a large number of objects used as electrical devices.

ELECTRIC POWER DEFINED

One final definition is needed. When we talk about rates of energy change in an electrical circuit, we deal with two notions of the term, *power*. The power provided by sources of electromotive force, emf's, is given by the formula derived as follows. EMF's provide potential differences that move charges. The energy provided by the emf can be calculated by looking at infinitesimal charges moved through the potential difference, $dU = dq\ V = i\ V\ dt$ $dU/dt = i\ V = \text{Power}$.

Thus the power provided by an emf is the current times the potential difference it maintains between its positive and negative poles. In any material or device that has electrical resistance, the rate of change of energy for current flowing through it is lost or rather transformed into thermal energy. To determine the relationship for this, note that Ohm's Law tells us that the potential difference across a material or device with resistance R is $V_R = iR$. Therefore, the power necessary to maintain current i going through the material or device is $P = i\ V_R = i^2R = V_R^2/R$.As stated before, this power is the rate at which electrical energy is turned into thermal energy in the resistance.

Chapter 2

Magnetic Forces

As with the case of the electric field, we hypothesize the existence of such a field as a way of explaining action at a distance in which objects exert forces on each other without direct mechanical contact. As it is easy to distinguish between gravitational and electrical forces (and hence the nature of the field produced was originally hypothesized as being quite different in nature: electric fields act only on electric charges, gravitational fields produce forces only on mass, etc.), so too were magnetic fields and electric fields thought to be very different. After all, the nature of the forces produced are very different as we will see. However, there is a very important reason for thinking, before we know very much about magnetic fields, that they are closely linked with electric fields. Although this could not be known before the atomic theory of matter was believed and the basic constituents of atoms discovered, magnetic forces act *only* on electric charges as there are no confirmed observations of magnetic charges. The fact that magnetic fields exert forces on electric charges at all is a clue that somehow electricity and magnetism share a deep bond. To understand that bond, we need to explore thoroughly the properties of the magnetic field.

MAGNETIC FORCE ON A CHARGED PARTICLE

Experimentally, to isolate the nature of the magnetic field, we need to start, as with the electric field, with the simplest thing you can act on, namely a particle. The behaviour of magnetic forces is best indicated by how a charged particle behaves in the presence of such a field

A charged particle moves in the presence of a magnetic field. If the particle stops its motion, the magnetic force on it is zero. This is true independent of the direction or magnitude of the magnetic field. The mathematical description of the force due to the magnetic field is $F = qv \times B$.This formula describes a force which is, by virtue of the cross-product, always perpendicular to v and B and has a magnitude equal to qvB sinq where q is the angle between v and B. This is a general result for any charged particle moving in any magnetic field. The circular path seen in the applet results because the force is always perpendicular to the velocity (and also to the B field direction). If the velocity magnitude and B are constant, then the path of the particle is a circle since the force, and hence the acceleration, of the particle is always perpendicular to its path and constant in magnitude. This fact is used to make two very practical devices of the 20th century: the mass spectrometer and the cyclotron. Both of these have a multitude of scientific and practical uses.

FORCES ON CURRENTS

In the earliest studies of magnetic fields, individual charged particles were not known. The properties of magnetic fields and forces were studied using the effects of fields on current distributions. If we have a bunch of charges flowing through a straight wire of length L, then the net force is

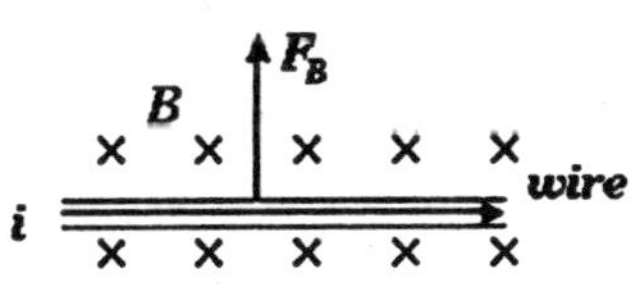

$$\vec{F} = \Sigma_q q\vec{v} \times \vec{B}$$

$$= \Sigma_q \frac{q\vec{L}}{t} \times \vec{B}$$

$$\vec{F} = i\vec{L} \times \vec{B}$$

Force on a wire carrying a current I in a magnetic field B.

where the direction of L is the same as the direction of the current.

We can apply the above formulas to the more general case of current distributions of arbitrary shape by breaking any line of current into infinitesimal lengths and integrating over them to find the net force, dF= idL × B

MAGNETIC FORCE ON AN ARBITRARILY SHAPED WIRE

We can generalize the above result to take into account any arbitrary shape of current distribution. For example, consider the following figure where some arbitrarily shaped wire takes a current i from point a to point b in a magnetic field B which has constant magnitude and direction perpendicular to the plane containing i.

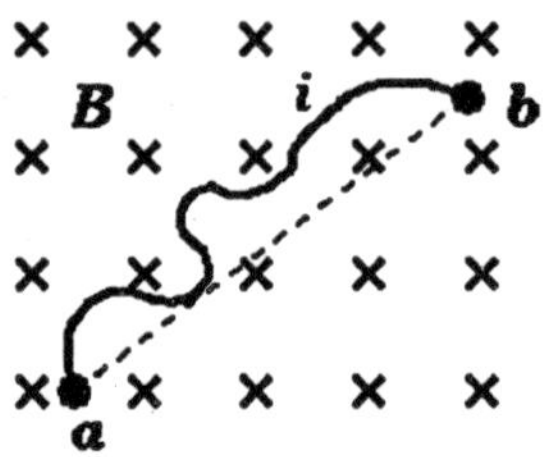

Fig. An arbitrarily shaped current distribution in an external magnetic field.

The problem is not as difficult as it might at first appear. Consider any particular piece of the wire.

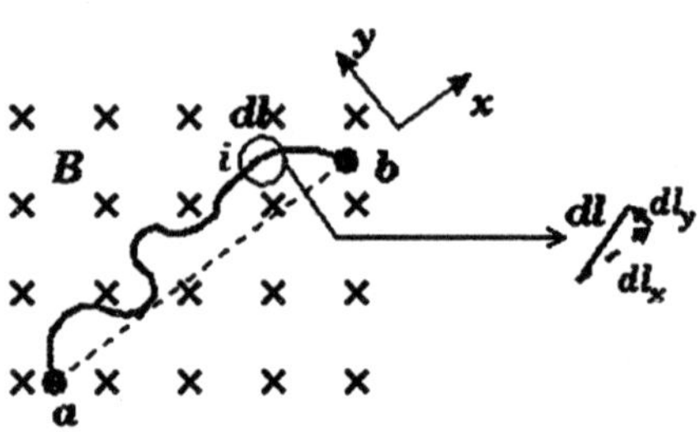

Fig. Blowup of a section of the wire.

In breaking this distribution into infinitesimal lengths, dl, if we note that we are free to do this such that dl has components that are both perpendicular and parallel to the line between points a and b (we called these the y and x axes, respectively), then note that the contributions on the force due to the y components will cancel. Why? Note that in going from a to b, our net displacement is only along x (remember: we have chosen the x axis to lie along the line between a and b). Hence, any +y displacement must eventually be countered with a -y displacement that cancels it. Since the magnetic force on any +y displacement is opposite in direction to that on any -y displacement, the integral over i^*dl_yB must be zero. Hence, we have a non-zero contribution only from the i^*dl_xB components. We note that the B field is perpendicular to both the x and y components. Hence our result is $F_{net} = iL_{ab}B$

where L_{ab} represents the distance from a to b along a straight line between the two points. If you go back to our previous example, you will see that this result is exactly consistent with this prediction.

MAGNETIC FIELD OF MOVING CHARGES

In keeping with our previous statement about the association of charges acted upon by fields and the fields produced by those charges, it is only natural to think that, once it is established that magnetic fields act on *moving* electric charges, then moving electric charges should create magnetic fields. In fact, experiment shows just such a relationship! The symmetry inherent in nature between fields and charges is therefore upheld. We find that a single moving point charge q which has constant velocity v in a vacuum produces a magnetic field

$$B = \mu_0/4\pi \times q\, v \times r/r^2$$

Note the introduction of a new fundamental constant of nature, m_0, the permeability of free space.

We *define* the value of m_0 as being

$$\mu_0 = 4p \times 10^{-7}\ N{\cdot}s^2/C^2 = 4\lambda \times 10^{-7}\ T{\cdot}m/A$$

So, if m_0 is a constant of nature, why is it defined rather than measured as other constants have been? The reason lies in what was historically recognized as a curious relationship between the permittivity and the permeability of free space. Namely, their product is the inverse of the square of the speed of light, i.e.

$$c^2 = 1/\varepsilon_0\mu_0$$

The reason for this amazing correspondence will be explained later in the semester. For now, we take it that it is useful to define the value of c and hence after measuring ε_0 we define the value of μ_0 to be consistent with the definition of c given the above correspondence.

BIOT-SAVART LAW

In determining the magnetic field for a bunch of moving charged particles which formulate a current, one must be guided by experiment. Starting from measurements of the magnetic field around a long (assume infinitely long) wire carrying a steady current, two French scientists, Jean Biot and Felix Savart, worked out the relationship relating magnetic field strength to the distance from the wire to the point where the field is measured as being 1/R with R being the distance. In keeping with nature's desire for symmetry, there are no preferred directions in space, hence all points on a circle of radius R and centred on the current distribution have the same field magnitude. It was found that the direction of the field is tangent to this circle at each point. Later, it was determined that the strength of the field was proportional to the current I in the wire. Hence, the magnitude and direction of the field look as shown in figure

$$B = \frac{\mu_0 I}{2\pi R}$$

Starting from experimental observation of the above, a theoretical formula can be obtained that reflects the contribution due to current through infinitesimal lengths of wire all summed together vectorially through integration. This formula, first worked out by Biot and Savart, is referred to as the Biot-Savart Law is the magnetic field contribution of a current element, i.e. an infinitesimal length d*l* of current I where d*l* has the direction of I and r is the vector from the current element to the position where the field is to be calculated.

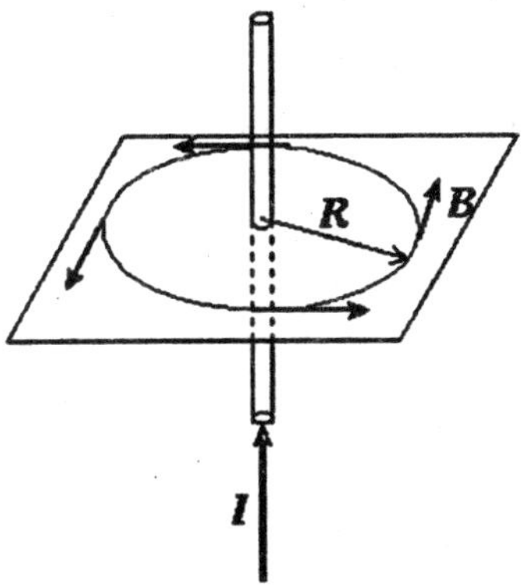

Fig. Magnetic field around a straight conductor carrying a current I.

Note that implicit in the Biot-Savart Law is the notion that magnetic fields obey the *principle of superposition*. This means that we can get the net field by just vectorially adding the contributions of the fields due to the infinite number of infinitesimal current elements.

Applying the Biot-Savart Law

To gain experience with applying this law, let's consider some specific examples, starting with the infinite straight wire carrying a steady current I.

DEFINITION OF AMPERE'S LAW

Andre Marie Ampere was not in agreement with Biot and Savart as to their experimental determination of the B field. First of all, isolated "current elements" such as appear in the law are difficult to justify "physically". Currents are always part of a complete circuit and hence identifying them as individual, isolated "elements" denies their true nature. In his own experimental and theoretical studies, Ampere determined an alternative law that can be shown to be in agreement with Biot and Savart's Law for the case of an infinitely long, infinitesimally thin wire.

Ampere's Law is stated as:

$$\oint \vec{B} \cdot d\vec{l} = \mu_0 I_{enc}$$

where I_{enc} is the current penetrating the area enclosed by the path over which we integrate. The obvious example to apply this to is the long, thin wire. Here, the wire is carrying current into the page and we wish to find the magnetic field at a point P a distance r away. As in all examples of using Ampere's Law, we wish to find a closed path, i.e. one that bounds an area, that allows us to separate B from the integral. To do this, we need a path in which B is constant in magnitude and parallel, antiparallel, or perpendicular to d*l* for every part of the path. The reasons for this are:

1. In order to be mathematically sensible, the path must define an area through which the current passes, i.e. a non-closed path does not define a definitive area through which we can unambiguously state what current is contributing to the B field along the path. Therefore, the path must be closed.
2. Parts of the path must be parallel (or anti-parallel) to the B field so that we have something to evaluate for B·d*l*. If, for part of the path, we have B perpendicular to d*l*, then, for those parts, B·d*l* = 0. In the end, we wish. to know the magnitude of B. Therefore, we must have B·d*l* be known over the whole path and known in such a way that *all* of B is known, not just a component of it.
3. Clearly, for this to work, you can only use Ampere's Law for cases in which you already know a lot about the behaviour of B. You need to know its direction and something about how its magnitude changes over the possible paths you might choose. In

addition, the B field must have enough symmetry to make it possible for you to complete steps 1 and 2 above.

Applying Ampere's Law to a Long, Thin Wire

The obvious path for the long, thin wire is shown below as the dashed line labeled d*l*.

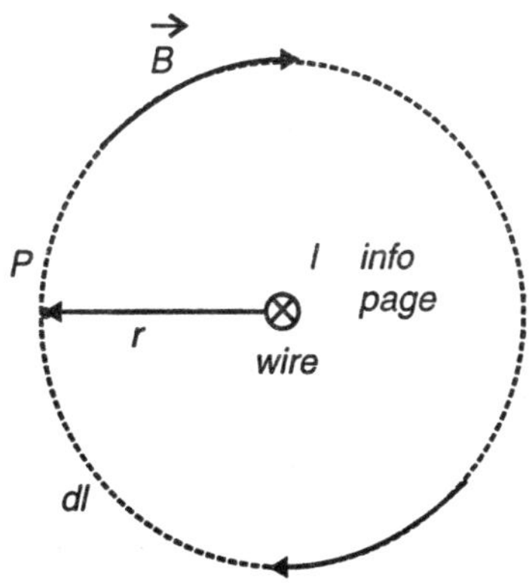

We use this circular path of radius r since we know, from the Biot-Savart Law, that the magnetic field around the wire is tangent to the circle in the clockwise direction and that it's magnitude depends only on r, not on position parallel to the wire. Thus, we use Ampere's Law to derive

$$B.dl = \mu_0 I_{enc}, \quad B = \mu_0 I/2\pi r$$

To be sure that Ampere's Law is consistent with the Biot-Savart Law, let's see if it works for some path other than the circular one we chose. For example, look at the path below which has two circular arc sections centred on the wire carrying current into the page. One arc has radius a and the other radius b. They are connected by straight sections.

Note that the path is closed and that we do know the characteristics of the B field. The circled numbers indicate that we have identified 4 pieces of the path for the application of Ampere's Law. These four pieces give

where we need to know what B_1 and B_3 are and we have used the fact that, for the straight paths 2 and 4, the B field is perpendicular to the path dl at every point along the path. The Biot-Savart Law also tells us that

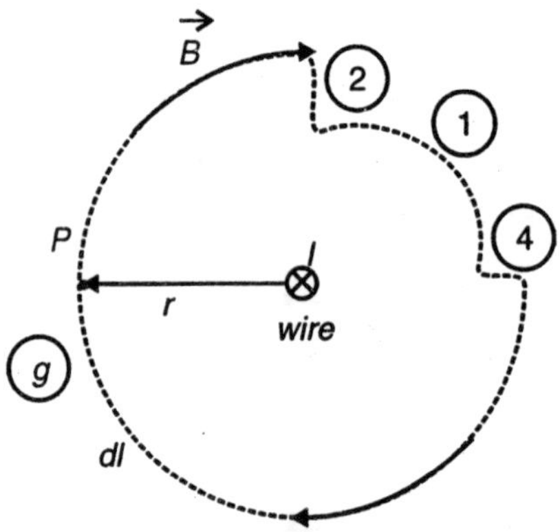

Fig. Ampere's Law applied to a straight current into the page and an Amperian path which is not circular.

$B = \mu_0 I/2\pi b$

If you plug these into our previous result, you find that Ampere's Law is verified. In practical terms, this means that Ampere's Law is always valid. It may not always be useful for finding the B field. In this respect, it is exactly like Gauss's Law in that you can only apply it to learn about the B field if you happen to recognize a symmetry of the field that allows you to bring B outside the integral of B·dl.

Magnetic Field of a Solenoid

We can readily find examples where Ampere's Law is more useful, by virtue of being easier to apply, than the Biot-Savart Law. For example, as was the case for the electric field, it was deemed to be very practical to have a device which could "store" magnetic field as a capacitor "stores" an electric field. We know that what the capacitor really stores is electric charge separation, but one of the properties for which a parallel-plate capacitor is useful is its ability to produce a uniform electric field between its plates. The equivalent device for magnetic fields is the solenoid. It's simply a conducting wire wrapped into a cylindrical shape. Even though the actual shape of the wire is helical, for densely packed wrapping we can actually consider the solenoid to be a bunch of closely spaced coils. Near the centre of each coil we know that the magnetic field is nearly perpendicular to the plane of the coil as shown in figure

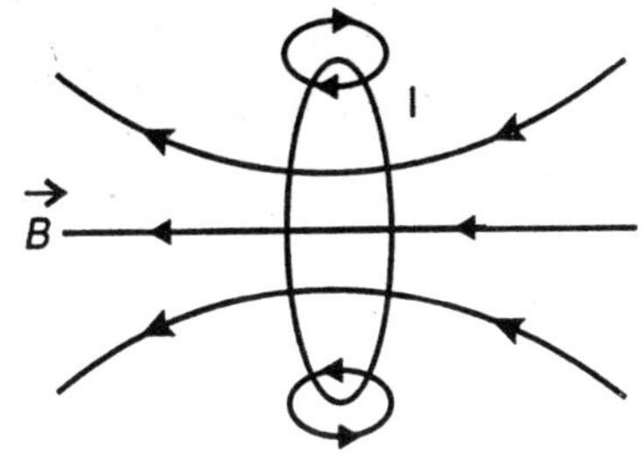

Fig. The magnetic field near the centre of a single coil carrying a current I.

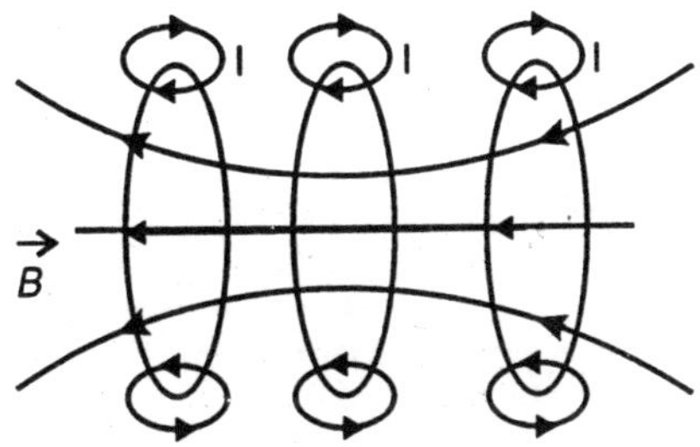

Fig. The magnetic field near the centre of a set of three coils all carrying a current I.

If we place coils on either side of the first and let them all carry the same current I in the same direction, then the magnetic field lines will link together near the centre to make a field that is approximately uniform in direction and strength near the centre of the set of coils. Placing many coils in close proximity to each other yields the solenoid field.

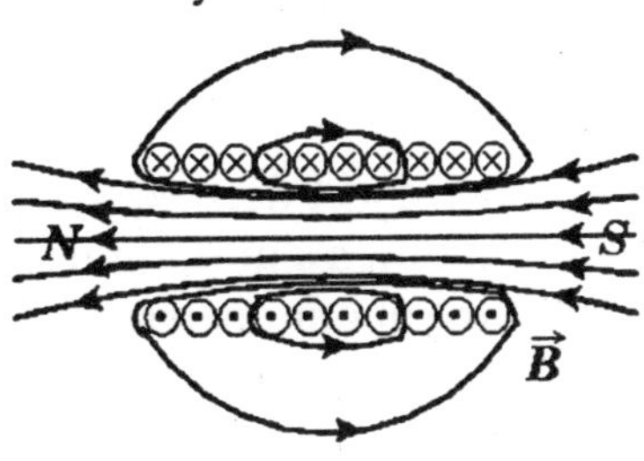

A solenoid approximates many current-carrying coils spaced much more closely than the coil diameter.

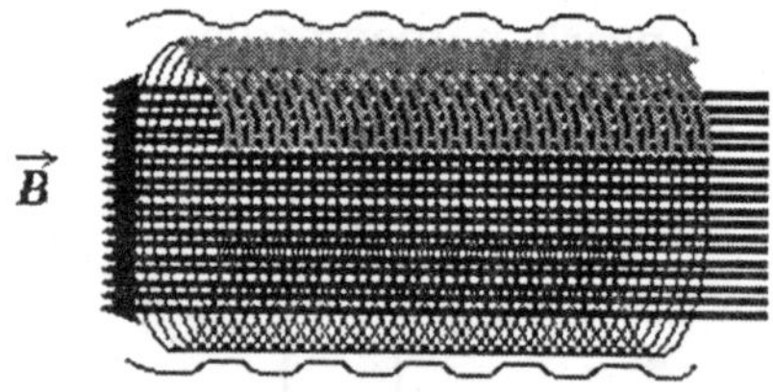

A perfect solenoid has coils so close together that the magnetic field is zero outside the solenoid and perfectly uniform inside.

Approximating the field as constant in direction and magnitude near the centre of the solenoid allows us to use Ampere's Law to calculate its magnitude.

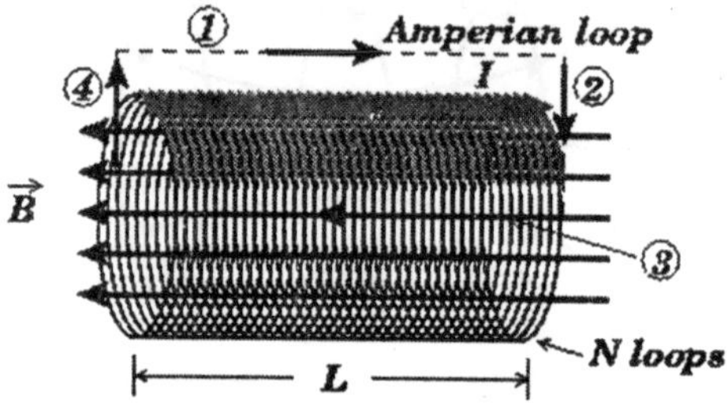

Fig. Applying Ampere's Law to a solenoid.

Choosing a flat rectangle as the Amperian loop, we see that the contributions to the loop integral can be broken into four parts. The area bounded by the flat rectangle is penetrated by N loops over a length L.Hence our result is *B.dl*= $B_{solenoid}L = m_0NI$ $B_{solenoid} = m_0nI$.where n = N/L is the linear density of current loops.

Magnetic Field of a Toroid

We can use what we just derived to examine the case of another device that produces a uniform magnetic field. In this case, we take advantage of our analysis of the solenoid to ask what happens if we bend a solenoid into a circle so that the ends join. The new configuration still has approximately zero field in the regions outside the volume contained by the coils, but the field inside that volume is again approximately uniform if the distance between the coils is small compared to the size of the coils. This device is a toroid.

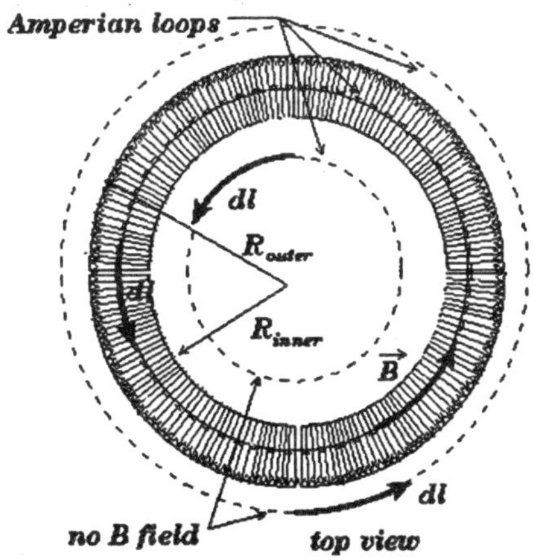

Ampere's Law applied to a toroid. Note that Amperian loops which lie wholly outside the volume contained by the toroidal coils experience no magnetic field.

We expect the magnetic field to have circular symmetry about the centre of the toroid using the same reasoning as for the solenoid. Hence, we expect it to be most useful to use circular paths for evaluating the Ampere integral of magnetic field and path. For any circular path whose area is not intersected by the coils, the magnetic field is zero and the current penetrating the area is, by definition, zero. Hence, $B_{r<R_{inner}} = 0$. For a circular path whose radius is greater than R_{outer}, we also expect to have no magnetic field. This might seem to be a contradiction with Ampere's Law because the current in the toroid coils definitely penetrate the area contained, however, note that the coils take current through the area (out of the page) at R_{inner} and *back into* the page at R_{outer}. So, the net current penetrating the area contained by the circular loop is zero since every loop of the wire carries current through

the area twice - in opposite directions! So it is consistent that the magnetic field be zero.

Magnetic Field of an Infinite Sheet of Current

We can consider an instance of a current distribution which does not have cylindrical symmetry, but which is susceptible to Ampere's Law for finding the magnitude of the magnetic field. Consider a sheet of current which is infinitesimally thin but infinitely long and wide. The sheet has a linear current density (i.e. current per unit length), λ.

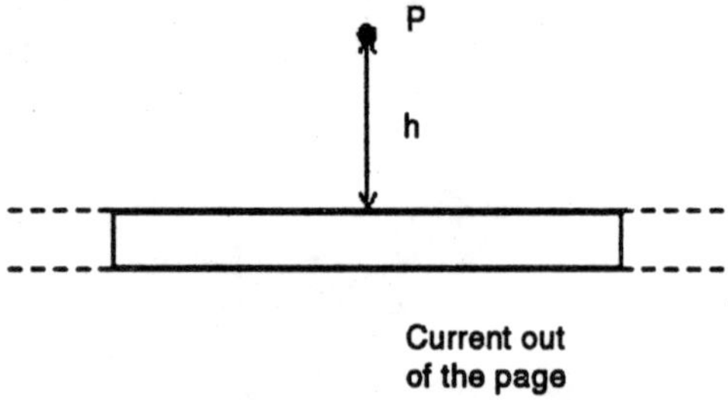

Fig. An infinite sheet of current with current per unit length λ. We wish to find the magnetic field direction and magnitude at point P a distance h away from the sheet.

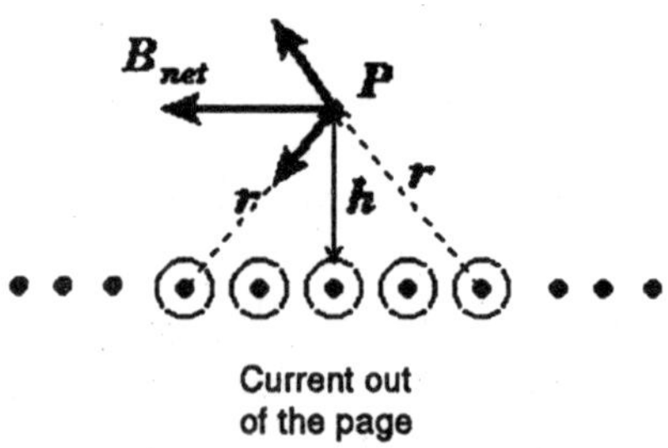

Fig. Consider a set of wires laid in place of the current sheet. Each wire carries current out of the page. The magnetic field due to wires which are equidistant from the point directly underneath point P can only be in the horizontal direction.

We can replace the sheet with an infinite set of wires arranged so that each wire carries current consistent with an overall current per unit length of l as in the case of the current sheet. We see in figure that wires which are equidistant from the line from the set of wires to point P have magnetic fields whose vertical components cancel and whose horizontal components add. Hence, the net magnetic field is in the horizontal direction. This field is uniform since the distribution of wires is infinite, i.e. any position for point P can be considered as the "middle" of an infinitely long current sheet. Note also that the magnetic field direction for the infinite sheet is independent of the distance of P from the sheet and that the same arguments for extending consideration of a finite set of wires to an infinite current sheet state that the magnetic field direction on the other side of the sheet has the field pointing in the opposite direction. Now we can apply Ampere's Law for the path shown in figure . Now we can apply Ampere's Law for the path shown in the figure.

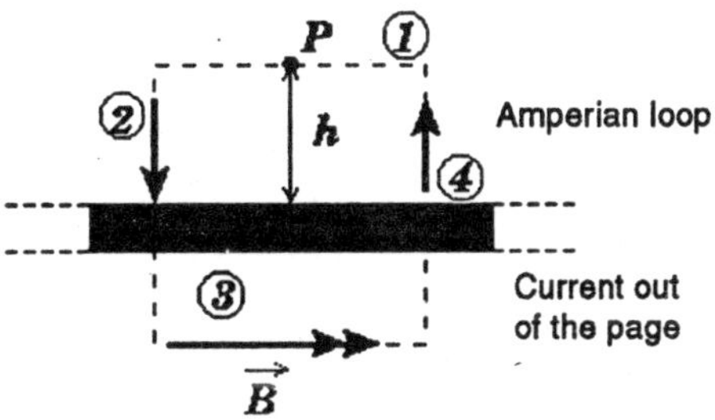

Fig. The Amperian loop for the infinite current sheet.

The loop gives us $B.dl = \mu_0 I_{enc.}$, $B_1 \cdot dl + B_2 \cdot dl + B_3 \cdot dl + B_4 \cdot dl = m_0 \lambda L$

Notice that the magnetic field is always perpendicular to parts 2 and 4 of the path. Part 3 is along the field direction below the sheet since, as stated before, the same symmetry arguments work here to say that the field should be parallel to the sheet, but in the opposite direction to the field on the other side of the sheet. Therefore,

$$B_1 \cdot dl + B_2 \cdot dl + B_3 \cdot dl + B_4 \cdot dl = \mu_0 \lambda L$$

$$B_P L + 0 + B_P L + 0 = \lambda_0 \lambda L$$

$$B_P = \tfrac{1}{2} \mu_0 \lambda$$

FARADAY'S LAW

Symmetry in nature predicts that, if a magnetic field acting on a loop of conductor carrying a current can produce a torque on the loop (assuming that the magnetic moment of the loop is not aligned with the field), then a torque acting on a conducting loop in a magnetic field should produce a current.

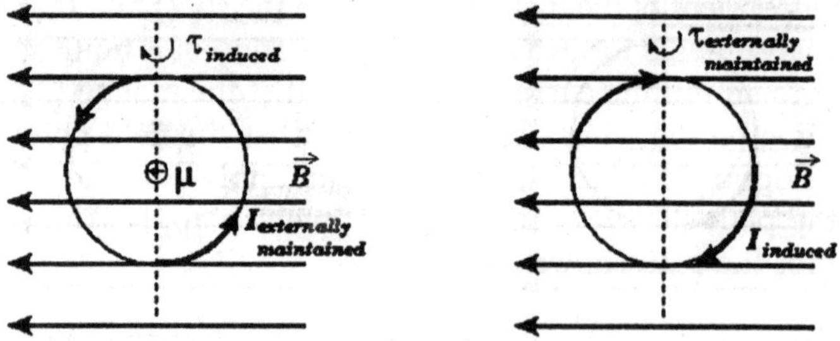

Symmetry in producing a torque for a loop with an externally maintained current going through it is balanced in nature by having an externally applied torque on a conducting loop produced a current in the opposite direction.

Note that the symmetry cannot be "perfect" in that the direction of the current that is induced in the loop due to the externally maintained torque must be opposite to the direction of current that would create a torque in that direction. This is necessary due to energy conservation. If the effect happened as in our first case (externally maintained current), then more current would go through the loop and the induced torque would increase. This violates energy conservation as the change in angular position due to the external torque would induce still more change in angular position. Hence we note that if the effect of induced current does take place, the current must be induced in a direction that *opposes* the external torque that causes it.

Faraday was the first to show that this induced current effect does happen. In studying the effect, we find that the magnitude and direction of the induced current effect is due to a change in the magnetic flux through an area in space. This effect is referred to as electromagnetic induction and the quantitative description of the effect is Faraday's Law. Specifically, Faraday's Law states that $\varepsilon = -d\phi_B/dt$.

In other words, Faraday expressed his effect by saying that the magnitude of the EMF induced by any change of magnetic flux is proportional to the magnitude of the change in magnetic flux. The induced EMF is what causes the current to flow. To correspond to the necessity of energy conservation, the direction of the EMF must be such that the induced current would oppose the change in the magnetic flux. This was best expressed by H.F.E. Lenz in Lenz's Law.

The Direction of any Magnetic Induction Effect must Oppose the Cause of the Effect.

The easiest way to generate a static picture of this is to observe the four cases below of a bar magnet being pushed toward and away from a conducting loop as shown in figure. The current created by the induced EMF must oppose the change in each case.

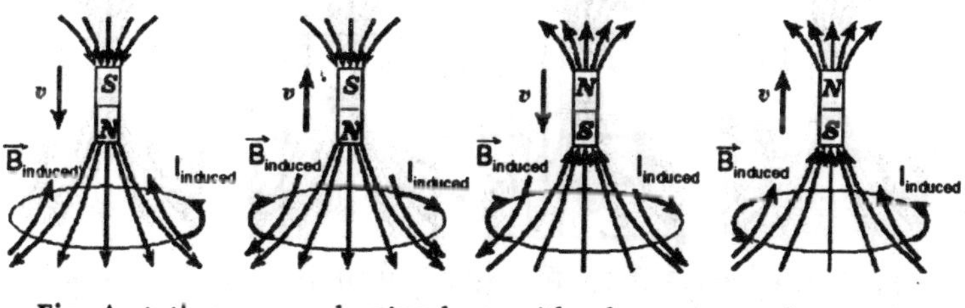

Fig. A stationary conducting loop with a bar magnet that can be alternately pushed toward or away from the loop with either of two orientations of the poles (i.e. North pole toward the loop or away from it).

Make sure you understand figure. It shows, in each case, how the induced magnetic field from the induced current (which comes from the induced EMF assuming the loop has some amount of resistance) opposes the change in magnetic flux. Since magnetic flux is a scalar, it might be difficult to think of a "direction" for it, but the sign of the flux depends on the arbitrary convention for the sign for area, A. Once that is set, then the dot product of B and A gives the positive or negative sign for the flux. The negative sign in Faraday's Law then gives the appropriate sign for the direction of the induced EMF. In the first case of figure, the induced current creates an induced field which points opposite to the external magnetic field from the bar magnet since the flux going down through the loop is increasing with time. For the third case, the induced current creates an induced field which points opposite to the external magnetic field since the flux pointing up through the loop is increasing. Verify that the second and fourth cases shown make sense to you following the same argument. The importance of Faraday's effect is hard to overestimate. Most of the world runs on electricity generated as a result of it.

Faraday's Law Applied

If we consider the case for a conducting bar sliding over rails in the presence of a magnetic field, we have a good case for examining both Lenz's Law and considering a quantitative example of Faraday's Law. Figure shows the situation.

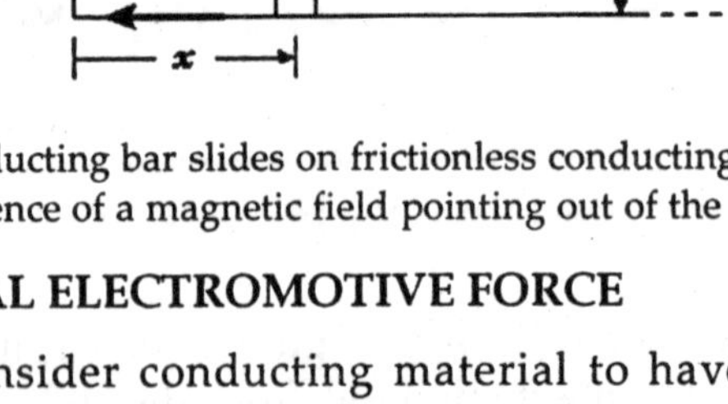

Fig. A conducting bar slides on frictionless conducting rails in the presence of a magnetic field pointing out of the page.

MOTIONAL ELECTROMOTIVE FORCE

We consider conducting material to have a number (essentially infinite) number of electric charge carriers that are free to move (albeit with resistance unless the conductor is a perfect conductor) throughout the material in any way demanded by external electric and magnetic fields. That means we can consider the source of an induced EMF by looking at the direct effects of an external magnetic field on charges in a moving conductor. For example, if we consider a conducting bar of length L moving in a plane perpendicular to a constant magnetic field, we see that the positive and negative electric charges within the bar will separate as shown in figure.

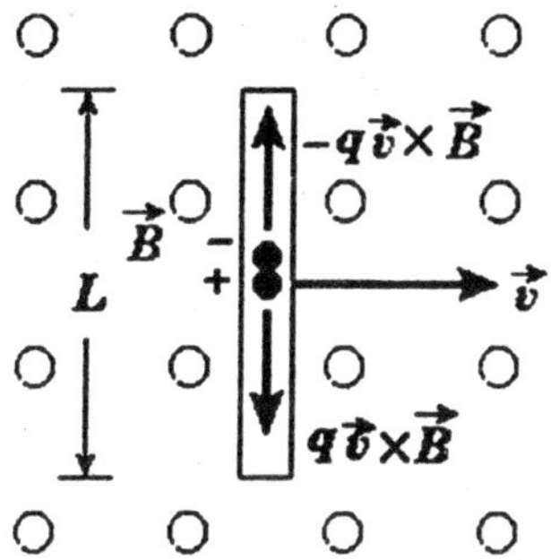

Fig. A conducting bar moving in a magnetic field experiences separation of positive and negative charges within the bar. This leads to an induced EMF across the bar.

The positive charges are pushed to the bottom by the velocity cross magnetic field force and the negative charges are pushed in the opposite direction. The charge separation creates an electric field in the downward direction. This electric field builds as the charge separation grows until the force on the charges due to the resulting electric field cancels the force due to motion in the magnetic field. That means that we have, if we assume the charge q to mean positive charge, $e_{induced}$ = vBL

where the motional EMF, ε, is just the average electric field divided by the length of the bar. It is induced by the motion of the bar in the magnetic field. The magnitude of this EMF is equal in magnitude to the product of the velocity, the magnetic field magnitude, and the length of the bar. If we attach this bar to frictionless rails which provide a resistance R as shown in figure and provide an external force on the bar to maintain its constant velocity, then the bar is a *source* of EMF and just like any other EMF source, can drive a current through the circuit.

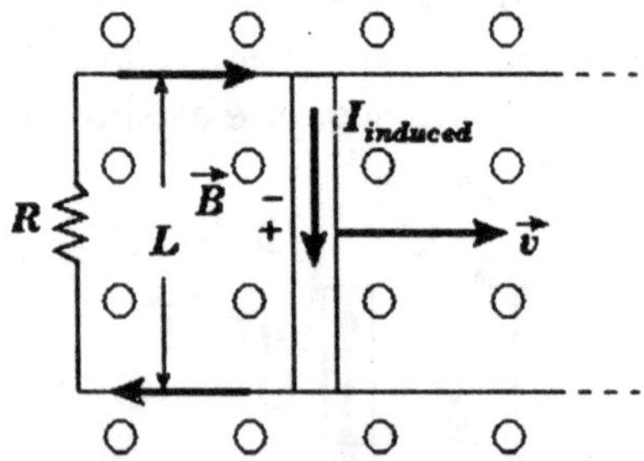

Fig. A conducting bar moving on frictionless rails acts as an EMF source for pushing current through the circuit formed by the bar and the rails.

Note that the motional EMF is identical to the induced EMF we calculated through Faraday's Law. In fact, motional EMF is one example of Faraday's Law. It has a number of practical applications.

Note that the direction of the EMF in both the top and bottom pieces of the loop point to the right. Hence the bottom EMF pushes current counter-clockwise about the loop and the top EMF pushes current clockwise as shown in figure.

The motional EMF on each of the four sides of the conducting loop can be calculated. The net EMF pushes current counter-clockwise around the loop.

The net EMF for the circuit is therefore $e_{net} = e_{bot} - e_{top} = B_0 l v_y (y_{bot} - y_{top}) = B_0 l w v_y$

if we arbitrarily assign counter-clockwise as the positive direction. The induced current in the loop is $I_{induced} = e_{net}/R = B_0 l w v_y/R$

INDUCED ELECTRIC FIELDS

In general, since the force on charge carriers in a conductor is given by F = qv × B,we can calculate the motional EMF for a conductor moving in a magnetic field via the following formulation.

$$E \cdot dl = \varepsilon = -dF_B/dt$$

Hence we see that, in general, we can calculate motional EMF's, i.e. the EMF generated in *moving conductors* through the interaction of the motion and external magnetic field. The second equation shows that Faraday's Law predicts the creation of an induced electric field in the presence of a time-varying magnetic flux. Note that this second equation and the original expression of Faraday's Law are equivalent and *always* valid. The calculation of motional EMF's is consistent with Faraday's Law *only* for the case of motion of a conductor in a magnetic field.

AMPERE-MAXWELL LAW

James Clerk Maxwell was responsible for setting down our current understanding of the fundamental importance of the equations that make up Gauss's Law, Faraday's Law, and Ampere's Law. In deciding on a suitable description of electrodynamics, he sought to place the mathematical description on the same footing as Newton's Laws of Motions. That he succeeded is testimony to his mathematical skill and great physical insight.

To start, we should review here the form of the laws as we currently describe them.

Gauss's Law for Static Electric Charges	$E \cdot dA = q/\varepsilon_0$
Gauss's Law for Magnetism	$B \cdot dA = 0$
Ampere's Law	$B \cdot dl = \mu_0 I_{enc.}$
Faraday's Law	$E \cdot dl = -d\phi_B/dt$

where the second equation represents the fact that there are no magnetic charges observed (so far) in the universe, hence there are no point-like sources or sinks of the magnetic field. We first note the degree of mathematical symmetry present in the equations despite the differences in behaviour of electric and magnetic fields. It's also clear that this symmetry between electric and magnetic fields would be perfect except for two obvious asymmetries within the equations themselves. The first is that there are no magnetic charges as there are electric charges (something that has motivated physicists to seek out such charges for decades) and the second is that, electric fields can be created in *two* ways: either from static electric charges (Gauss's Law) or induced by time-varying magnetic fields (magnetic fluxes to be exact), but there is no corresponding possibility of creating magnetic fields from either static magnetic charges (they have not been found to exist) nor from time-varying electric fields. If we simply assumed that the latter symmetry had to exist, why would it not be apparent in experiments that were done up to the 1800's? In this, we can make a small digression that points out one of the prime means of gaining insights into physics research.Let us infer that nature is respectful of symmetry mathematically. What would Maxwell's Equations look like completely symmetrically? First, the existence of magnetic charges would change the form of Gauss's Law applied to static charges. We would have $E \cdot dA = q_E/\varepsilon_{0,}$ $EB \cdot dA = \mu_0\, q_{B..}$ If units are to be kept consistent. What about currents involving electric and magnetic charges? Certainly there would be contributions to both Ampere's Law and Faraday's Law. These contributions, again looking at consistent units, would look as follows: $E{\cdot}dl = -d\phi_B/dt\ \mu_0 I_{B.}$ where the subscripts E and B depict electric or magnetic charges. The k in the revamped Ampere's Law is a placeholder for the correct proportionality constant for this term. We can find the value of this constant in terms of the fundamental constants m_0 and e_0 purely from dimensional analysis. Since magnetic monopoles do not exist, we deal simply with the additions necessary to consider induced magnetic fields from time-varying electric field fluxes. This would affect only Ampere's Law and the term added would assume that a time-varying electric flux would induce a B field. To make sure that the units are correct, we would want to have Ampere's Law extended as follows:

$B{\cdot}dl = \mu_0(I_{enc.} + I_{displacement}),$

where we want $I_{displacement} = d\phi_E/dt,$

That is, this new term $I_{displacement}$ is proportional to the time-rate of change of electric flux. The simplest form of converting electric flux (SI units of $N/(C \cdot m^2 \cdot s)$) to current

(C/s) is, interestingly enough, multiplication by a constant with the units $C^2/(N \cdot m^2)$. Note that these are precisely the units of the permittivity constant, $\varepsilon_{0.}$ Hence, $I_{displacment} = \varepsilon_0 \, d\Phi_E/dt$.

This comes directly from dimensional analysis and a "belief" in the need for symmetry among the equations for electricity and magnetism. Note that this ansatz also gives rise to an assumption as to why the effect might be hard to observe. Note that, if the above form is correct, then Ampere's Law is changed to have the form $B \cdot dl = \mu_0 I_{enc.} + \mu_0 \varepsilon_0 \, d\Phi_E/dt$

This means that the time-varying electric flux term is multiplied by *two* small terms rather than just one. This might make the effect weak enough to be masked by other experimental interferences unless one is very careful. In fact, the form we have derived for $I_{displacement}$, referred to as the displacement current, does exist and can be verified experimentally. It was motivated by Maxwell theoretically not from the symmetry and dimensional analysis argument above, but from the following consideration of a weakness in Ampere's mathematical description of his law. Consider the case of a capacitor connected to conducting lines. If one of the lines carries a current I, then the other line connected to the other plate must also carry such a current due to the electric field of the previous plate.

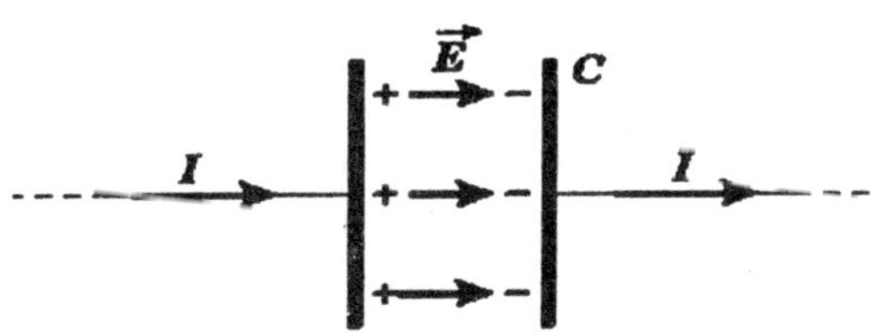

Fig. A charging capacitor and the corresponding time-varying electric field between its plates.

If we apply Ampere's Law to a circular disk around the wire, then the magnetic field is expected to be constant in magnitude and to have a direction which follows the boundary. The magnitude of the B field is proportional to the current penetrating the area bound by the path.

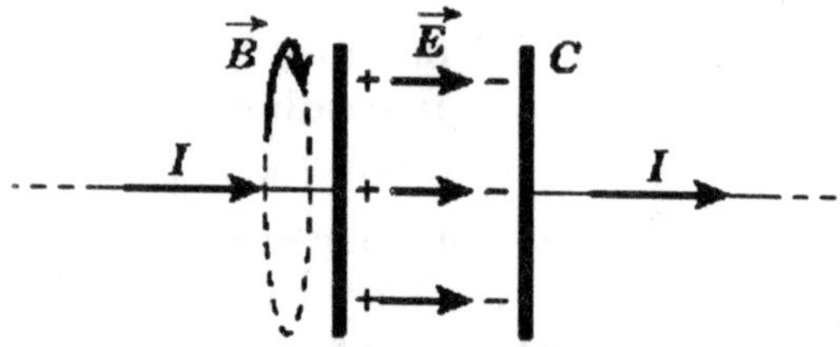

Fig. Magnetic field around a current as it approaches the capacitor.

Unfortunately, the mathematical description of the area bounded by a curve is rather loose. The area shown in figure is bounded by the same curve.

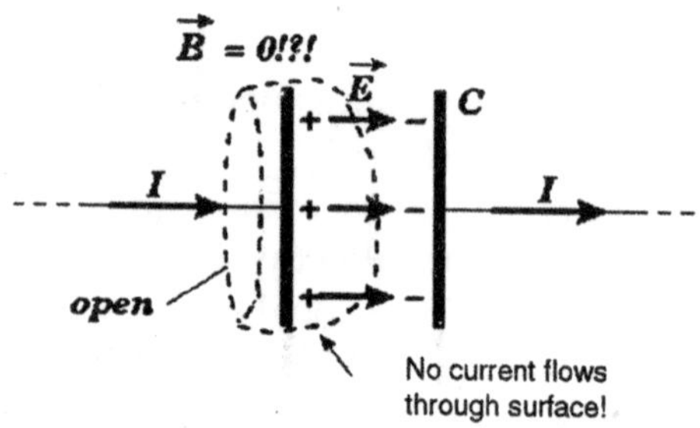

Fig. Magnetic field around the same perimeter which encloses a different area.

Ampere's Law, as stated, is a line integral which can be shown to be equivalent for the two areas and in fact for any area bounded by the same circular curve around the wire in the region outside the capacitor.The unfortunate part, though, is that no current penetrates the second surface and hence the prediction is that the magnetic field should be zero around the boundary.To resolve the obvious inconsistency, Maxwell generalized Ampere's Law to include the displacement current matching just the characteristics we gave on the basis of the dimensional argument above.

To verify the form of the displacement current, note that a parallel-plate capacitor builds charge according to the current impinging on the left plate for our figure. This current is related to the change in electric flux as follows: $\Phi_E/dt = I/\varepsilon_0.$ If we multiply this by e_0, we see that we get the displacement current as being equal to the current in the wires attached to the plates of the capacitors, hence we expect to get the same value for the magnetic field no matter whether we draw the surface so as to have the current in the wire penetrate the surface or to have the time-varying electric field between the capacitor plates penetrating the surface.

Chapter 3

Magnetism

Magnets are very common items in the workplace and household. Uses of magnets range from holding pictures on the refrigerator to causing torque in electric motors. Most people are familiar with the general properties of magnets but are less familiar with the source of magnetism. The traditional concept of magnetism centres around the magnetic field and what is know as a dipole. The term "magnetic field" simply describes a volume of space where there is a change in energy within that volume. This change in energy can be detected and measured. The location where a magnetic field can be detected exiting or entering a material is called a magnetic pole. Magnetic poles have never been detected in isolation but always occur in pairs, hence the name dipole. Therefore, a dipole is an object that has a magnetic pole on one end and a second, equal but opposite, magnetic pole on the other.

A bar magnet can be considered a dipole with a north pole at one end and south pole at the other. A magnetic field can be measured leaving the dipole at the north pole and returning the magnet at the south pole. If a magnet is cut in two, two magnets or dipoles are created out of one. This sectioning and creation of dipoles can continue to the atomic level. Therefore, the source of magnetism lies in the basic building block of all matter...the atom.

THE SOURCE OF MAGNETISM

All matter is composed of atoms, and atoms are composed of protons, neutrons and electrons. The protons and neutrons are located in the atom's nucleus and the electrons are in constant motion around the nucleus. Electrons carry a negative electrical charge and produce a magnetic field as they move through space. A magnetic field is produced whenever an electrical charge is in motion. The strength of this field is called the magnetic moment.

This may be hard to visualize on a subatomic scale but consider electric current flowing through a conductor. When the electrons (electric current) are flowing through the conductor, a magnetic field forms around the conductor. The magnetic field can be detected using a compass. The magnetic field will place a force on the compass needle, which is another example of a dipole.

Since all matter is comprised of atoms, all materials are affected in some way by a magnetic field. However, not all materials react the same way.

DIAMAGNETIC, PARAMAGNETIC, AND FERROMAGNETIC MATERIALS

When a material is placed within a magnetic field, the magnetic forces of the material's electrons will be affected. This effect is known as Faraday's Law of Magnetic Induction. However, materials can react quite differently to the presence of an external magnetic field. This reaction is dependent on a number of factors, such as the atomic and molecular structure of the material, and the net magnetic field associated with the atoms. The magnetic moments associated with atoms have three origins. These are the electron orbital motion, the change in orbital motion caused by an external magnetic field, and the spin of the electrons.

In most atoms, electrons occur in pairs. Electrons in a pair spin in opposite directions.So, when electrons are paired together, their opposite spins cause their magnetic fields to cancel each other. Therefore, no net magnetic field exists. Alternately, materials with some unpaired electrons will have a net magnetic field and will react more to an external field. Most materials can be classified as diamagnetic, paramagnetic or.ferromagnetic.

Diamagnetic metals have a very weak and negative susceptibility to magnetic fields. Diamagnetic materials are slightly repelled by a magnetic field and the material does not retain the magnetic properties when the external field is removed. Diamagnetic materials are solids with all paired electron resulting in no permanent net magnetic moment per atom. Diamagnetic properties arise from the realignment of the electron orbits under the influence of an external magnetic field. Most elements in the periodic table, including copper, silver, and gold, are diamagnetic.

Paramagnetic metals have a small and positive susceptibility to magnetic fields. These materials are slightly attracted by a magnetic field and the material does not retain the magnetic properties when the external field is removed. Paramagnetic properties are due to the presence of some unpaired electrons, and from the realignment of the electron orbits caused by the external magnetic field. Paramagnetic materials include magnesium, molybdenum, lithium, and tantalum.

Ferromagnetic materials have a large and positive susceptibility to an external magnetic field. They exhibit a strong attraction to magnetic fields and are able to retain their magnetic properties after the external field has been removed. Ferromagnetic materials have some unpaired electrons so their atoms have a net magnetic moment. They get their strong magnetic properties due to the presence of magnetic domains. In these domains, large numbers of atom's moments (10^{12} to 10^{15}) are aligned parallel so that the magnetic force within the domain is strong. When a ferromagnetic material is in the unmagnitized state, the domains are nearly randomly organized and the net magnetic field for the part as a whole is zero. When a magnetizing force is applied, the domains become aligned to produce a strong magnetic field within the part. Iron, nickel, and cobalt are examples of ferromagnetic materials. Components with these materials are commonly inspected using the magnetic particle method.

MAGNETIC DOMAINS

Ferromagnetic materials get their magnetic properties not only because their atoms carry a magnetic moment but also because the material is made up of small regions known as magnetic domains. In each domain, all of the atomic dipoles are coupled together in a preferential direction. This alignment develops as the material develops its crystalline structure during solidification from the molten state. Magnetic domains can be detected using Magnetic Force Microscopy (MFM) and images of the domains like the one shown below can be constructed.

During solidification, a trillion or more atom moments are aligned parallel so that the magnetic force within the domain is strong in one direction. Ferromagnetic materials are said to be characterized by "spontaneous magnetization" since they obtain saturation magnetization in each of the domains without an external magnetic field being applied. Even though the domains are magnetically saturated, the bulk material may not show any signs of magnetism because the domains develop themselves and are randomly oriented relative to each other.

Ferromagnetic materials become magnetized when the magnetic domains within the material are aligned. This can be done by placing the material in a strong external magnetic field or by passing electrical current through the material. Some or all of the domains can become aligned. The more domains that are aligned, the stronger the magnetic field in the material. When all of the domains are aligned, the material is said to be magnetically saturated. When a material is magnetically saturated, no additional amount of external magnetization force will cause an increase in its internal level of magnetization.

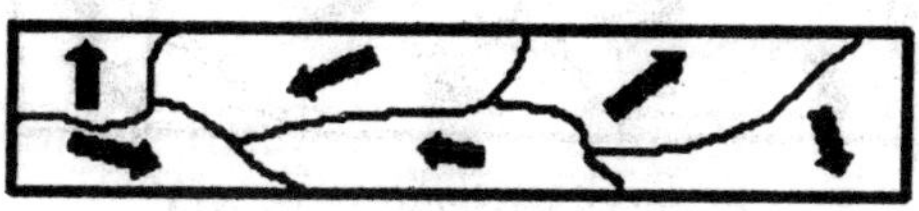

Fig. Unmagnetized Material

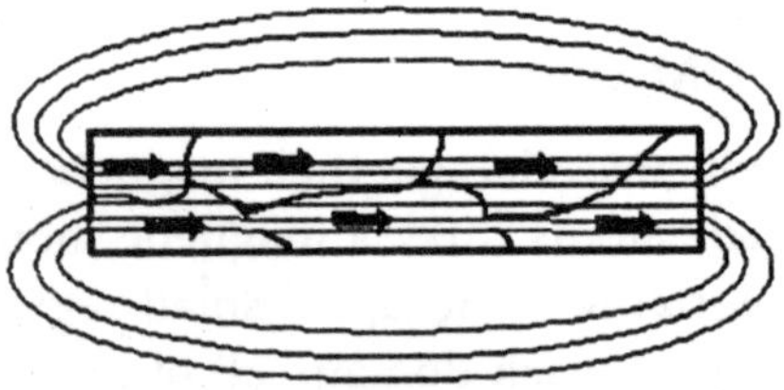

Fig. Magnetized Material

MAGNETIC FIELD CHARACTERISTICS

As discussed previously, a magnetic field is a change in energy within a volume of space. The magnetic field surrounding a bar magnet can be seen in the magnetograph below. A magnetograph can be created by placing a piece of paper over a magnet and sprinkling the paper with iron filings. The particles align themselves with the lines of magnetic force produced by the magnet. The magnetic lines of force show where the magnetic field exits the material at one pole and reenters the material at another pole along the length of the magnet. It should be noted that the magnetic lines of force exist in three dimensions but are only seen in two dimensions in the image.

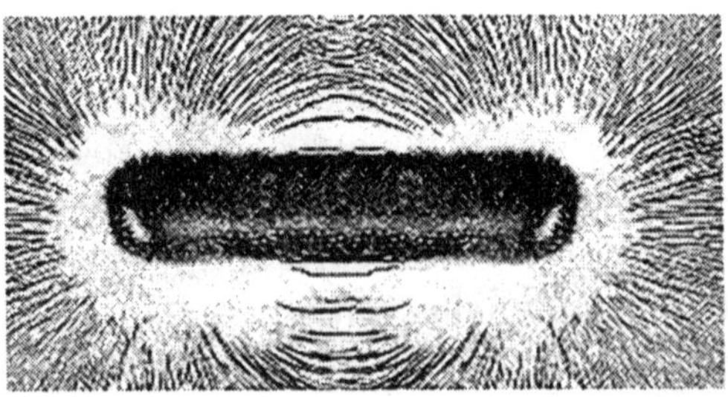

It can be seen in the magnetograph that there are poles all along the length of the magnet but that the poles are concentrated at the ends of the magnet. The area where the exit poles are concentrated is called the magnet's north pole and the area where the entrance poles are concentrated is called the magnet's south pole.

Magnetic Fields in and around Horseshoe and Ring Magnets.

Magnets come in a variety of shapes and one of the more common is the horseshoe (U) magnet.The horseshoe magnet has north and south poles just like a bar magnet but the magnet is curved so the poles lie in the same plane. The magnetic lines of force flow from pole to pole just like in the bar magnet. However, since the poles are located closer together and a more direct path exists for the lines of flux to travel between the poles, the magnetic field is concentrated between the poles.

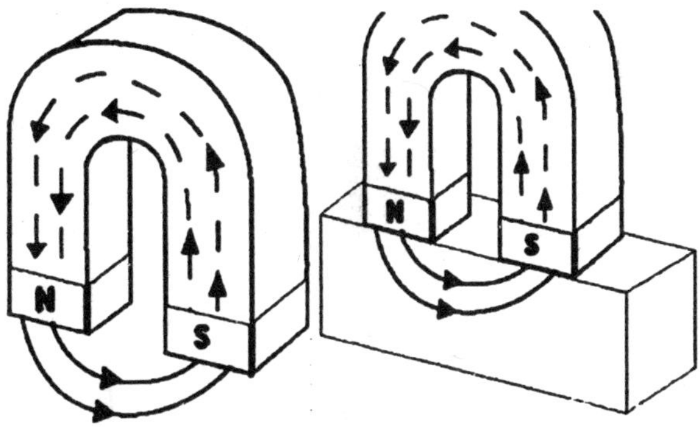

If a bar magnet was placed across the end of a horseshoe magnet or if a magnet was formed in the shape of a ring, the lines of magnetic force would not even need to enter the air. The value of such a magnet where the magnetic field is completely contained with the material probably has limited use. However, it is important to understand that the magnetic field can flow in loop within a material.

General Properties of Magnetic Lines of Force

Magnetic lines of force have a number of important properties, which include:

- They seek the path of least resistance between opposite magnetic poles. In a single bar magnet as shown to the right, they attempt to form closed loops from pole to pole.
- They never cross one another.
- They all have the same strength.

- Their density decreases (they spread out) when they move from an area of higher permeability to an area of lower permeability.
- Their density decreases with increasing distance from the poles.
- They are considered to have direction as if flowing, though no actual movement occurs.
- They flow from the south pole to the north pole within a material and north pole to south pole in air.

ELECTROMAGNETIC FIELDS

Magnets are not the only source of magnetic fields. In 1820, Hans Christian Oersted discovered that an electric current flowing through a wire caused a nearby compass to deflect. This indicated that the current in the wire was generating a magnetic field. Oersted studied the nature of the magnetic field around the long straight wire. He found that the magnetic field existed in circular form around the wire and that the intensity of the field was directly proportional to the amount of current carried by the wire. He also found that the strength of the field was strongest next to the wire and diminished with distance from the conductor until it could no longer be detected. In most conductors, the magnetic field exists only as long as the current is flowing (i.e. an electrical charge is in motion). However, in ferromagnetic materials the electric current will cause some or all of the magnetic domains to align and a residual magnetic field will remain.

Oersted also noticed that the direction of the magnetic field was dependent on the direction of the electrical current in the wire. A three-dimensional representation of the magnetic field is shown below. There is a simple rule for remembering the direction of the magnetic field around a conductor. It is called the right-hand rule. If a person grasps a conductor in one's right hand with the thumb pointing in the direction of the current, the fingers will circle the conductor in the direction of the magnetic field.

A Word of Caution about the right-hand Rule

For the right-hand rule to work, one important thing that must be remembered about the direction of current flow. Standard convention has current flowing from the positive terminal to the negative terminal. This convention is credited to Benjamin Franklin who theorized that electric current was due to a positive charge moving from the positive terminal to the negative terminal. However, it was later discovered that it is the movement of the negatively charged electron that is responsible for electrical current. Rather than changing several centuries of theory and equations, Franklin's convention is still used today.

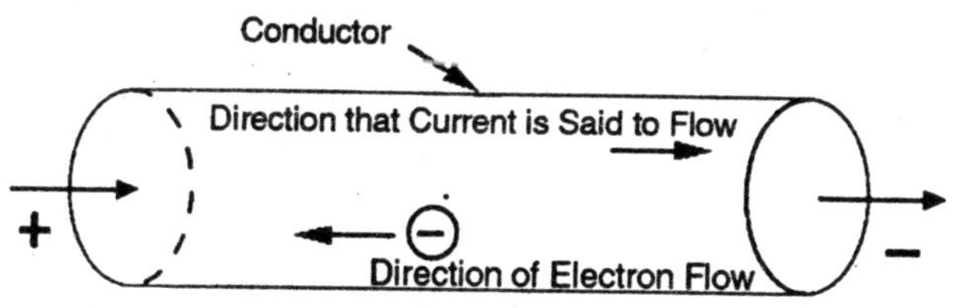

MAGNETIC FIELD PRODUCED BY A COIL

When a current carrying conductor is formed into a loop or several loops to form a coil, a magnetic field develops that flows through the centre of the loop or coil along its longitudinal axis and circles back around the outside of the loop or coil. The magnetic field circling each loop of wire combines with the fields from the other loops to produce a concentrated field down the centre of the coil. A loosely wound coil is illustrated below to show the interaction of the magnetic field. The magnetic field is essentially uniform down the length of the coil when it is wound tighter.

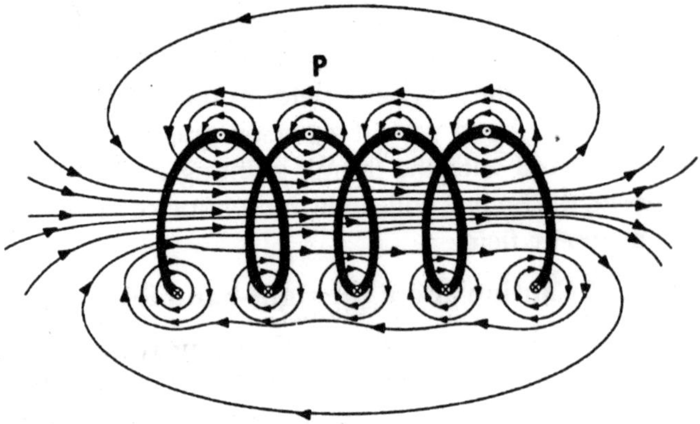

The strength of a coil's magnetic field increases not only with increasing current but also with each loop that is added to the coil. A long, straight coil of wire is called a solenoid and can be used to generate a nearly uniform magnetic field similar to that of a bar magnet. The concentrated magnetic field inside a coil is very useful in magnetizing ferromagnetic materials for inspection using the magnetic particle testing method. Please be aware that the field outside the coil is weak and is not suitable for magnetizing ferromagnetic materials.

QUANTIFYING MAGNETIC PROPERTIES

Until now, only the qualitative features of the magnetic field have been discussed. However, it is necessary to be able to measure and express quantitatively the various characteristics of magnetism. Unfortunately, a number of unit conventions are used. SI units will be used in this material. The advantage of using SI units is that they are traceable back to an agreed set of four base units - meter, kilogram, second, and Ampere.

Quantity		SI Units (Sommerfeld)	SI Units (Kennelly)	CGS Units (Gaussian)
Field	H	A/m	A/m	oersteds
Flux Density (Magnetic Induction)	B	tesla	tesla	gauss
Flux	ϕ	weber	weber	maxwell
Magnetization	M	A/m	-	erg/Oe-cm^3

The units for magnetic field strength H are ampere/meter. A magnetic field strength of 1 ampere/meter is produced at the centre of a single circular conductor with a one meter diameter carrying a steady current of 1 ampere.

The number of magnetic lines of force cutting through a plane of a given area at a right angle is known as the magnetic flux density, B. The flux density or magnetic induction has the tesla as its unit. One tesla is equal to 1 Newton/(A/m). From these units, it can be seen that the flux density is a measure of the force applied to a particle by the magnetic field. The Gauss is the CGS unit for flux density and is commonly used by US industry. One gauss represents one line of flux passing through one square centimeter of air oriented 90 degrees to the flux flow.

The total number of lines of magnetic force in a material is called magnetic flux, ϕ. The strength of the flux is determined by the number of magnetic domains that are aligned within a material. The total flux is simply the flux density applied over an area. Flux carries the unit of a weber, which is simply a tesla- meter^2.

The magnetization is a measure of the extent to which an object is magnetized. It is a measure of the magnetic dipole moment per unit volume of the object. Magnetization carries the same units as a magnetic field: amperes/meter.

Table: Conversion between CGS and SI magnetic units.

Quantity	CGS unit	SI Unit
Magnetic field	H = 10e	H=1000/4π A.m–1
Magnetic induction	B = 1 gauss	B = 1 × 10–4 Tesla
Magnetization	M = 1erg.Oe–1.cm–3	M = 1 × 103 A.m–1

THE HYSTERESIS LOOP AND MAGNETIC PROPERTIES

A great deal of information can be learned about the magnetic properties of a material by studying its hysteresis loop. A hysteresis loop shows the relationship between the induced magnetic flux density (B) and the magnetizing force (H). It is often referred to as the B-H loop. An example hysteresis loop is shown below.

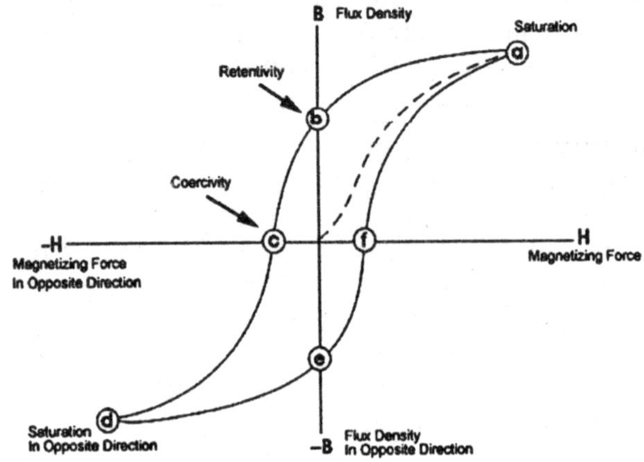

The loop is generated by measuring the magnetic flux of a ferromagnetic material while the magnetizing force is changed. A ferromagnetic material that has never been previously magnetized or has been thoroughly demagnetized will follow the dashed line as H is increased. As the line demonstrates, the greater the amount of current applied (H+), the stronger the magnetic field in the component (B+). At point "a" almost all of the magnetic domains are aligned and an additional increase in the magnetizing force will produce very little increase in magnetic flux. The material has reached the point of magnetic saturation. When H is reduced to zero, the curve will move from point "a" to point "b." At this point, it can be seen that some magnetic flux remains in the material even though the magnetizing force is zero. This is referred to as the point of retentivity on the graph and indicates the remanence or level of residual magnetism in the material.

(Some of the magnetic domains remain aligned but some have lost their alignment.) As the magnetizing force is reversed, the curve moves to point "c", where the flux has been reduced to zero. This is called the point of coercivity on the curve. (The reversed magnetizing force has flipped enough of the domains so that the net flux within the material is zero.) The force required to remove the residual magnetism from the material is called the coercive force or coercivity of the material.

As the magnetizing force is increased in the negative direction, the material will again become magnetically saturated but in the opposite direction (point "d"). Reducing H to zero brings the curve to point "e." It will have a level of residual magnetism equal to that achieved in the other direction. Increasing H back in the positive direction will return B to zero. Notice that the curve did not return to the origin of the graph because some force is required to remove the residual magnetism. The curve will take a different path from point "f" back to the saturation point where it with complete the loop.

From the hysteresis loop, a number of primary magnetic properties of a material can be determined.

1. *Retentivity*: A measure of the residual flux density corresponding to the saturation induction of a magnetic material. In other words, it is a material's ability to retain a certain amount of residual magnetic field when the magnetizing force is removed after achieving saturation. (The value of B at point b on the hysteresis curve.)
2. *Residual Magnetism or Residual Flux*: The magnetic flux density that remains in a material when the magnetizing force is zero. Note that residual magnetism and retentivity are the same when the material has been magnetized to the saturation point. However, the level of residual magnetism may be lower than the retentivity value when the magnetizing force did not reach the saturation level.

3. *Coercive Force*: The amount of reverse magnetic field which must be applied to a magnetic material to make the magnetic flux return to zero. (The value of H at point c on the hysteresis curve.)
4. *Permeability, m*: A property of a material that describes the ease with which a magnetic flux is established in the component.
5. *Reluctance*: Is the opposition that a ferromagnetic material shows to the establishment of a magnetic field. Reluctance is analogous to the resistance in an electrical circuit.

PERMEABILITY

As previously mentioned, permeability is a material property that describes the ease with which a magnetic flux is established in a component. It is the ratio of the flux density to the magnetizing force and is represented by the following equation:

$$\mu = B/H$$

It is clear that this equation describes the slope of the curve at any point on the hysteresis loop. The permeability value given in papers and reference materials is usually the maximum permeability or the maximum relative permeability. The maximum permeability is the point where the slope of the B/H curve for the unmagnetized material is the greatest. This point is often taken as the point where a straight line from the origin is tangent to the B/H curve.

The relative permeability is arrived at by taking the ratio of the material's permeability to the permeability in free space(air).

$$\mu(\text{relative}) = \mu(\text{material}) / \mu(\text{air})$$

$$\text{where: } \mu(\text{air}) = 1.256 \times 10^{-6} \text{ H/m}$$

The shape of the hysteresis loop tells a great deal about the material being magnetized. The hysteresis curves of two different materials are shown in the graph.

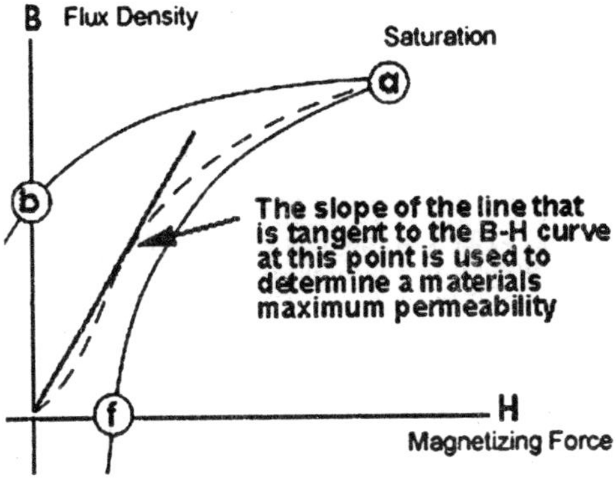

Relative to other materials, a material with a wider hysteresis loop has:

- Lower Permeability
- Higher Retentivity
- Higher Coercivity
- Higher Reluctance
- Higher Residual Magnetism

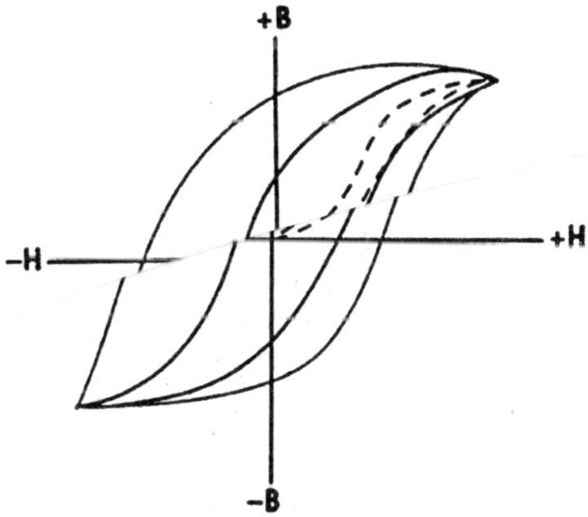

Relative to other materials, a material with the narrower hysteresis loop has:

- Higher Permeability
- Lower Retentivity

- Lower Coercivity
- Lower Reluctance
- Lower Residual Magnetism.

In magnetic particle testing, the level of residual magnetism is important. Residual magnetic fields are affected by the permeability, which can be related to the carbon content and alloying of the material. A component with high carbon content will have low permeability and will retain more magnetic flux than a material with low carbon content.

MAGNETIC FIELD ORIENTATION AND FLAW DETECTABILITY

To properly inspect a component for cracks or other defects, it is important to understand that the orientation between the magnetic lines of force and the flaw is very important. There are two general types of magnetic fields that can be established within a component.

A longitudinal magnetic field has magnetic lines of force that run parallel to the long axis of the part. Longitudinal magnetization of a component can be accomplished using the longitudinal field set up by a coil or solenoid. It can also be accomplished using permanent magnets or electromagnets.

A circular magnetic field has magnetic lines of force that run circumferentially around the perimeter of a part. A circular magnetic field is induced in an article by either passing current through the component or by passing current through a conductor surrounded by the component.

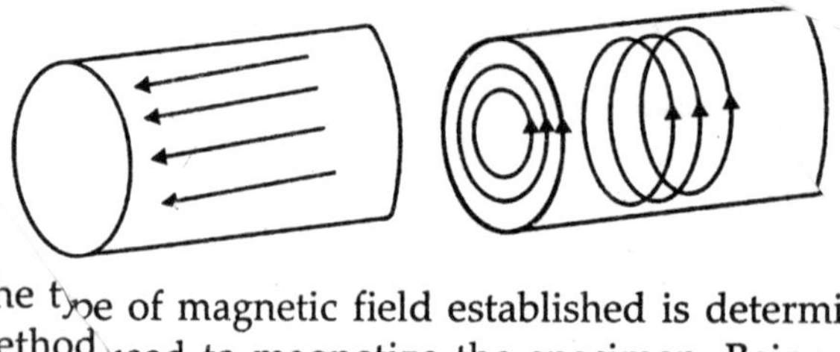

The type of magnetic field established is determined by the method used to magnetize the specimen. Being able to magnetize the part in two directions is important because the best detection of defects occurs when the lines of magnetic force are established at right angles to the longest dimension of the defect. This orientation creates the largest disruption of the magnetic field within the part and the greatest flux leakage at the surface of the part. As can be seen in the image, if the magnetic field is parallel to the defect, the field will see little disruption and no flux leakage field will be produced.

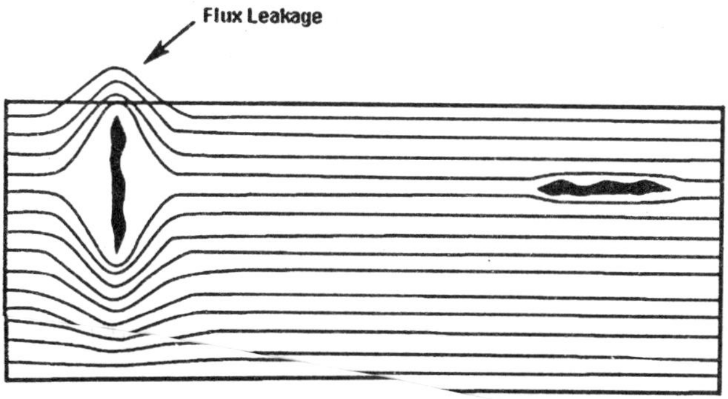

An orientation of 45 to 90 degrees between the magnetic field and the defect is necessary to form an indication. Since defects may occur in various and unknown directions, each part is normally magnetized in two directions at right angles to each other. If the component below is considered, it is known that passing current through the part from end to end will establish a circular magnetic field that will be 90 degrees to the direction of the current. Therefore, defects that have a significant dimension in the direction of the current (longitudinal defects) should be detectable. Alternately, transverse-type defects will not be detectable with circular magnetization.

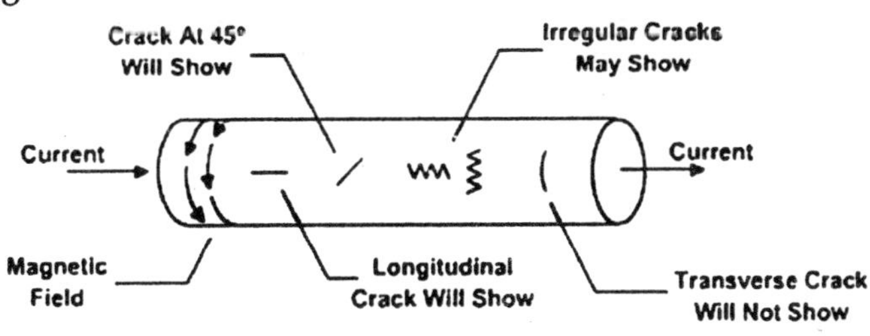

Watch this short movie showing the effect of field direction on indication visibility

MAGNETIZATION OF FERROMAGNETIC MATERIALS

There are a variety of methods that can be used to establish a magnetic field in a component for evaluation using magnetic particle inspection. It is common to classify the magnetizing methods as either direct or indirect.

Magnetization Using Direct Induction

With direct magnetization, current is passed directly through the component. Recall that whenever current flows, a magnetic field is produced. Using the right-hand rule, which was introduced earlier, it is known that the magnetic lines of flux form normal to the direction of the current and form a circular field in and around the conductor. When using the direct magnetization method, care must be taken to ensure that good electrical contact is established and maintained between the test equipment and the test component. Improper contact can result in arcing that may damage the component. It is also possible to overheat components in areas of high resistance such as the contact points and in areas of small cross-sectional area.

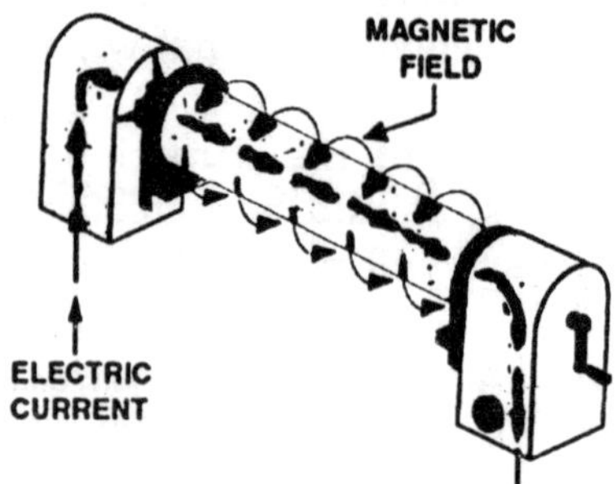

There are several ways that direct magnetization is commonly accomplished. One way involves clamping the component between two electrical contacts in a special piece of equipment. Current is passed through the component and a circular magnetic field is established in and around the component. When the magnetizing current is stopped, a residual magnetic field will remain within the component. The strength of the induced magnetic field is proportional to the amount of current passed through the component.

A second technique involves using clamps or prods, which are attached or placed in contact with the component. Electrical current flows through the component from contact to contact. The current sets up a circular magnetic field around the path of the current.

Magnetization Using Indirect Induction (Indirect Magnetization)

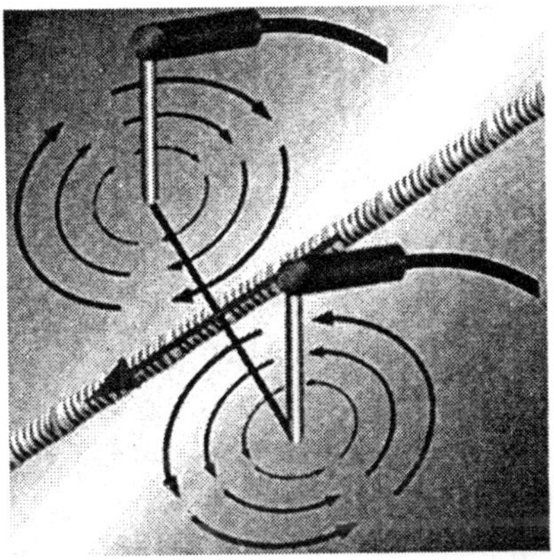

Indirect magnetization is accomplished by using a strong external magnetic field to establish a magnetic field within the component. As with direct magnetization, there are several ways that indirect magnetization can be accomplished.

The use of permanent magnets is a low cost method of establishing a magnetic field. However, their use is limited due to lack of control of the field strength and the difficulty of placing and removing strong permanent magnets from the component.

Electromagnets in the form of an adjustable horseshoe magnet (called a yoke) eliminate the problems associated with permanent magnets and are used extensively in industry. Electromagnets only exhibit a magnetic flux when electric current is flowing around the soft iron core. When the magnet is placed on the component, a magnetic field is established between the north and south poles of the magnet.

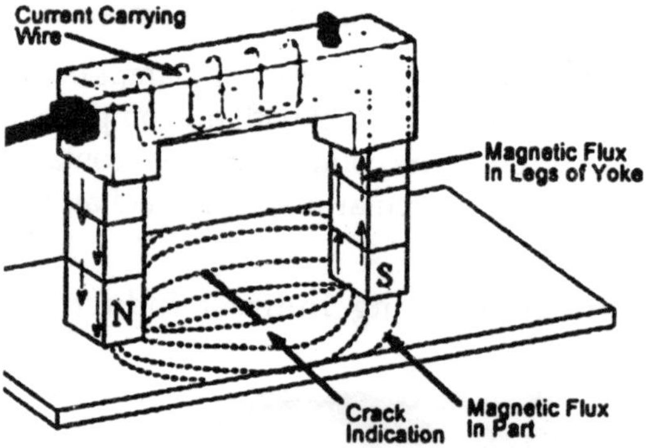

Another way of indirectly inducting a magnetic field in a material is by using the magnetic field of a current carrying conductor. A circular magnetic field can be established in cylindrical components by using a central conductor. Typically, one or more cylindrical components are hung from a solid copper bar running through the inside diameter. Current is passed through the copper bar and the resulting circular magnetic field establishes a magnetic field within the test components.

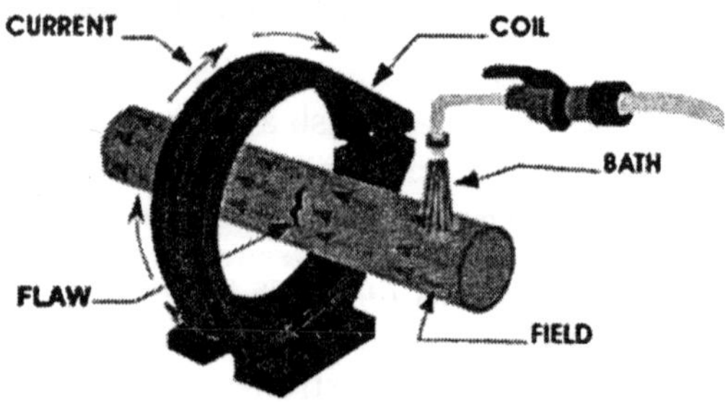

The use of coils and solenoids is a third method of indirect magnetization. When the length of a component is several times larger than its diameter, a longitudinal magnetic field can be established in the component. The component is placed longitudinally in the concentrated magnetic field that fills the centre of a coil or solenoid. This magnetization technique is often referred to as a "coil shot.

MAGNETIZING CURRENT

As seen in the previous pages, electric current is often used to establish the magnetic field in components during magnetic particle inspection. Alternating current and direct current are the two basic types of current commonly used. Current from single phase 110 volts, to three phase 440 volts, are used when generating an electric field in a component. Current flow is often modified to provide the appropriate field within the part. The type of current used can have an effect on the inspection results, so the types of currents commonly used will be briefly reviewed.

Direct Current

Direct current (DC) flows continuously in one direction at a constant voltage. A battery is the most common source of direct current. As previously mentioned, current is said to flow from the positive to the negative terminal. In actuality, the electrons flow in the opposite direction. DC is very desirable when inspecting for subsurface defects because DC generates a magnetic field that penetrates deeper into the material. In ferromagnetic materials, the magnetic field produced by DC generally penetrates the entire cross-section of the component. Conversely, the field produced using alternating current is concentrated in a thin layer at the surface of the component.

Alternating Current

Alternating current (AC) reverses in direction at a rate of 50 or 60 cycles per second. In the United States, 60 cycle current is the commercial norm but 50 cycle current is common in many countries. Since AC is readily available in most facilities, it is convenient to make use of it for magnetic particle inspection. However, when AC is used to induce a magnetic field in ferromagnetic materials, the magnetic field will be limited to narrow region at the surface of the component. This phenomenon is known as the "skin effect" and occurs because induction is not a spontaneous reaction and the rapidly reversing current does not allow the domains below the surface time to align. Therefore, it is recommended that AC be used only when the inspection is limited to surface defects.

Rectified Alternating Current

Clearly, the skin effect limits the use of AC since many inspection applications call for the detection of subsurface defects. However, the convenient access to AC, drives its use beyond surface flaw inspections. Luckily, AC can be converted to current that is very much like DC through the process of rectification. With the use of rectifiers, the reversing AC can be converted to a one directional current. The three commonly used types of rectified current are described below.

Half Wave Rectified Alternating Current (HWAC)

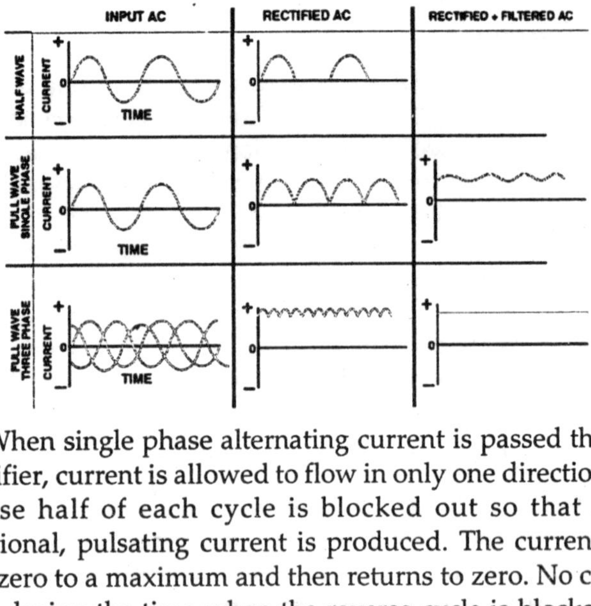

When single phase alternating current is passed through a rectifier, current is allowed to flow in only one direction. The reverse half of each cycle is blocked out so that a one directional, pulsating current is produced. The current rises from zero to a maximum and then returns to zero. No current flows during the time when the reverse cycle is blocked out. The HWAC repeats at same rate as the unrectified current (60 hertz typical). Since half of the current is blocked out, the amperage is half of the unaltered AC.

This type of current is often referred to as half wave DC or pulsating DC. The pulsation of the HWAC helps magnetic particle indications form by vibrating the particles and giving them added mobility. This added mobility is especially important when using dry particles. The pulsation is reported to significantly improve inspection sensitivity. HWAC is most often used to power electromagnetic yokes.

Full Wave Rectified Alternating Current (FWAC) (Single Phase)

Full wave rectification inverts the negative current to positive current rather than blocking it out. This produces a pulsating DC with no interval between the pulses. Filtering is usually performed to soften the sharp polarity switching in the rectified current. While particle mobility is not as good as half-wave AC due to the reduction in pulsation, the depth of the subsurface magnetic field is improved.

Three Phase Full Wave Rectified Alternating Current

Three phase current is often used to power industrial equipment because it has more favorable power transmission and line loading characteristics. This type of electrical current is also highly desirable for magnetic particle testing because when it is rectified and filtered, the resulting current very closely resembles direct current. Stationary magnetic particle equipment wired with three phase AC will usually have the ability to magnetize with AC or DC (three phase full wave rectified), providing the inspector with the advantages of each current form.

Longitudinal Magnetic Fields Distribution and Intensity

When the length of a component is several times larger than its diameter, a longitudinal magnetic field can be established in the component. The component is often placed longitudinally in the concentrated magnetic field that fills the centre of a coil or solenoid. This magnetization technique is often referred to as a "coil shot."

The magnetic field travels through the component from end to end with some flux loss along its length as shown in the image to the right. Keep in mind that the magnetic lines of flux occur in three dimensions and are only shown in 2D in the image. The magnetic lines of flux are much denser inside the ferromagnetic material than in air because ferromagnetic materials have much higher permeability than does air. When the concentrated flux within the material comes to the air at the end of the component, it must spread out since the air can not support as many lines of flux per unit volume. To keep from crossing as they spread out, some of the magnetic lines of flux are forced out the side of the component.

When a component is magnetized along its complete length, the flux loss is small along its length. Therefore, when a component is uniform in cross section and magnetic permeability, the flux density will be relatively uniform throughout the component. Flaws that run normal to the magnetic lines of flux will disturb the flux lines and often cause a leakage field at the surface of the component.

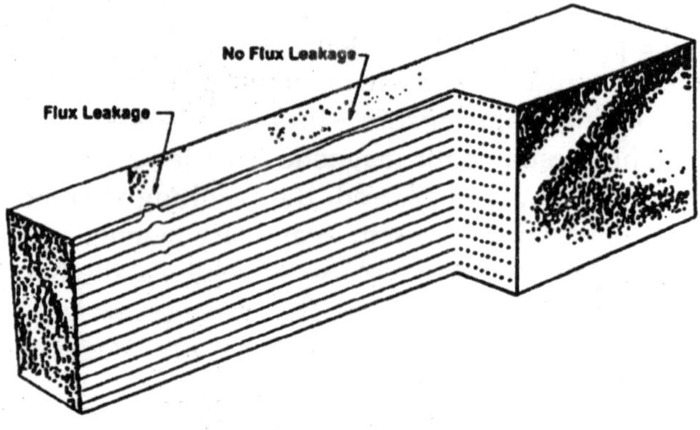

When a component with considerable length is magnetized using a solenoid, it is possible to magnetize only a portion of the component. Only the material within the solenoid and about the same width on each side of the solenoid will be strongly magnetized. At some distance from the solenoid, the magnetic lines of force will abandon their longitudinal direction, leave the part at a pole on one side of the solenoid and return to the part at a opposite pole on the other side of the solenoid. This occurs because the magnetizing force diminishes with increasing distance from the solenoid. As a result, the magnetizing force may only be strong enough to align the magnetic domains within and very near the solenoid. The unmagnetized portion of the component will not support as much magnetic flux as the magnetized portion and some of the flux will be forced out of the part as illustrated in the image. Therefore, a long component must be magnetized and inspected at several locations along its length for complete inspection coverage.

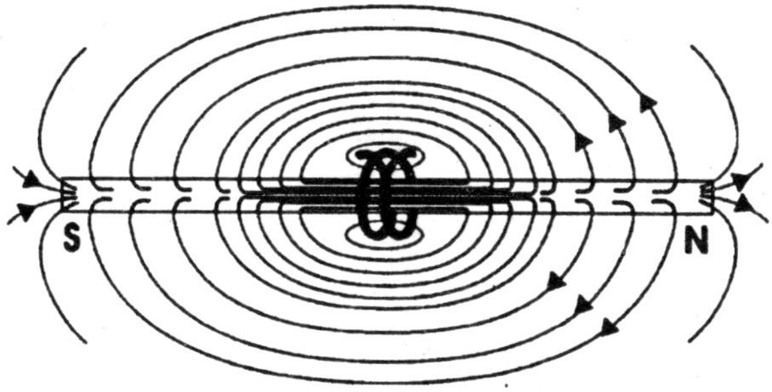

Solenoid - An electrically energized coil of insulated wire, which produces a magnetic field within the coil.

Circular Magnetic Fields Distribution and Intensity

As discussed previously, when current is passed through a solid conductor, a magnetic field forms in and around the conductor. The following statements can be made about the distribution and intensity of the magnetic field.

- The field strength varies from zero at the centre of the component to a maximum at the surface.
- The field strength at the surface of the conductor decreases as the radius of the conductor increases when the current strength is held constant. (However, a larger conductor is capable of carrying more current.)
- The field strength outside the conductor is directly proportional to the current strength. Inside the conductor, the field strength is dependent on the current strength, magnetic permeability of the material, and if magnetic, the location on the B-H curve.
- The field strength outside the conductor decreases with distance from the conductor.

In the images the magnetic field strength is graphed versus distance from the centre of the conductor. It can be seen that in a nonmagnetic conductor carrying DC, the internal field strength rises from zero at the centre to a maximum value at the surface of the conductor. The external field strength decrease with distance from the surface of the conductor. When the conductor is a magnetic material, the field strength within the conductor is much greater than it was in the nonmagnetic conductor. This is due to the permeability of the magnetic material. The external field is exactly the same for the two materials provided the current level and conductor radius are the same.

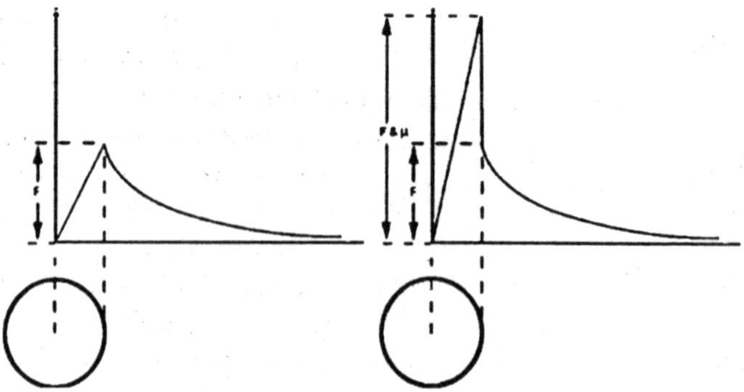

Fig. First:The magnetic field distribution in and around a solid conductor of a nonmagnetic material carrying direct current. The magnetic field distribution in and around a solid conductor of a magnetic material carrying direct current. Second: The magnetic field distribution in and around a solid conductor of a magnetic material carrying alternating current.

When the conductor is carrying alternating current, the internal magnetic field strength rises from zero at the centre to a maximum at the surface. However, the field is concentrated in a thin layer near the surface of the conductor. This is known as the "skin effect." The skin effect is evident in the field strength versus distance graph for a magnetic conductor shown to the right. The external field decreases with increasing distance from the surface as it does with DC. It should be remembered that with AC the field is constantly varying in strength and direction.

In a hollow circular conductor there is no magnetic field in the void area. The magnetic field is zero at the inside wall surface and rises until it reaches a maximum at the outside wall surface. As with a solid conductor, when the conductor is a magnetic material, the field strength within the conductor is much greater than it was in the nonmagnetic conductor due to the permeability of the magnetic material. The external field strength decreases with distance from the surface of the conductor. The external field is exactly the same for the two materials provided the current level and conductor radius are the same.

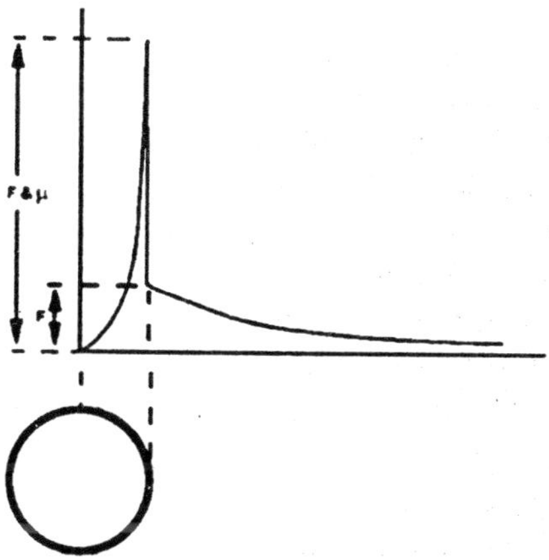

Fig. The magnetic field distribution in and around a hollow conductor of a magnetic material carrying alternating current.

The magnetic field distribution in and around a hollow conductor of a nonmagnetic material carrying direct current. The magnetic field distribution in and around a hollow conductor of a magnetic material carrying direct current.

When AC is passed through a hollow circular conductor, the skin effect concentrates the magnetic field at the outside diameter of the component.

As can be seen in the field distribution images, the field strength at the inside surface of hollow conductor carrying a circular magnetic field produced by direct magnetization is very low.

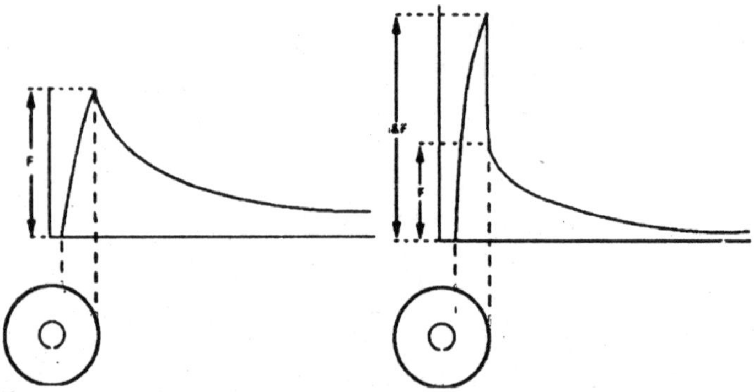

Fig. The magnetic field distribution in and around a nonmagnetic central conductor carrying DC inside a hollow conductor of a magnetic material.

Therefore, the direct method of magnetization is not recommended when inspecting the inside diameter wall of a hollow component for shallow defects.

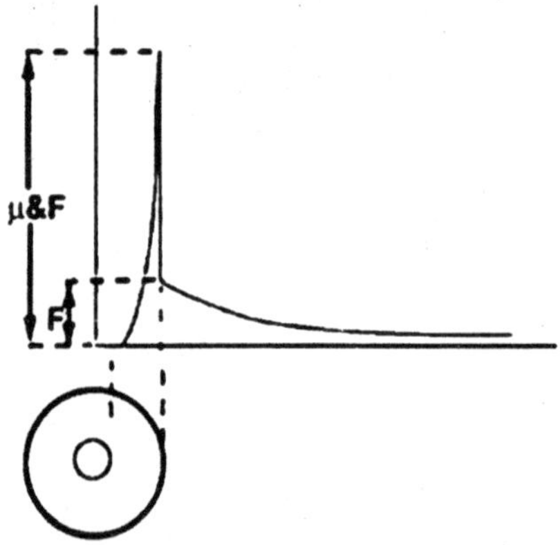

The field strength increases rather rapidly as one moves in from the ID, so if the defect has significant depth, it may be detectable. However, a much better method of magnetizing hollow components for inspection of the ID and OD surfaces is with the use of a central conductor. As can be seen in the field distribution image to the right, when current is passed through a nonmagnetic central conductor (copper bar), the magnetic field produced on the inside diameter surface of a magnetic tube is much greater and the field is still strong enough for defect detection on the OD surface.

DEMAGNETIZATION

After conducting a magnetic particle inspection, it is usually necessary to demagnetize the component. Remanent magnetic fields can:

- Affect machining by causing cuttings to cling to a component.
- Interfere with electronic equipment such as a compass.
- Create a condition known as "arc blow" in the welding process. Arc blow may cause the weld arc to wonder or filler metal to be repelled from the weld.
- Cause abrasive particles to cling to bearing or faying surfaces and increase wear.

Removal of a field may be accomplished in several ways. This random orientation of the magnetic domains can be achieved most effectively by heating the material above its curie temperature. The curie temperature for a low carbon steel is 770°C or 1390°F. When steel is heated above its curie temperature, it will become austenitic and loses its magnetic properties. When it is cooled back down, it will go through a reverse transformation and will contain no residual magnetic field. The material should also be placed with it long axis in an east-west orientation to avoid any influence of the Earth's magnetic field.

It is often inconvenient to heat a material above it curie temperature to demagnetize it, so another method that returns the material to a nearly unmagnetized state is commonly used. Subjecting the component to a reversing and decreasing magnetic field will return the dipoles to a nearly random orientation throughout the material. This can be accomplished by pulling a component out and away from a coil with AC passing through it. The same can also be accomplished using an electromagnetic yoke with AC selected. Also, many stationary magnetic particle inspection units come with a demagnetization feature that slowly reduces the AC in a coil in which the component is placed.

A field meter is often used to verify that the residual flux has been removed from a component. Industry standards usually require that the magnetic flux be reduced to less than 3 gauss after completing a magnetic particle inspection.

MEASURING MAGNETIC FIELDS

When performing a magnetic particle inspection, it is very important to be able to determine the direction and intensity of the magnetic field. As discussed previously, the direction of the magnetic field should be between 45 and 90 degrees to the longest dimension of the flaw for best detectability. The field intensity must be high enough to cause an indication to form, but not too high to cause nonrelevant indications to mask relevant indications. To cause an indication to form, the field strength in the object must produce a flux leakage field that is strong enough to hold the magnetic particles in place over a discontinuity. Flux measurement devices can provide important information about the field strength.

Since it is impractical to measure the actual field strength within the material, all the devices measure the magnetic field that is outside of the material. There are a number of different devices that can be used to detect and measure an external magnetic field. The two devices commonly used in magnetic particle inspection are the field indicator and the Hall-effect meter, which is also called a gauss meter. Pie gauges and shims are devices that are often used to provide an indication of the field direction and strength but do not actually yield a quantitative measure. They will be discussed in a later section.

Field Indicators

Field indicators are small mechanical devices that utilize a soft iron vane that is deflected by a magnetic field. The X-ray image shows the inside working of a field meter looking in from the side. The vane is attached to a needle that rotates and moves the pointer for the scale. Field indicators can be adjusted and calibrated so that quantitative information can be obtained. However, the measurement range of field indicators is usually small due to the mechanics of the device.

The one shown to the right has a range from plus 20 gauss to minus 20 gauss. This limited range makes them best suited for measuring the residual magnetic field after demagnetization.

Hall-Effect (Gauss/Tesla) Meter

A Hall-effect meter is an electronic device that provides a digital readout of the magnetic field strength in gauss or tesla units. The meters use a very small conductor or semiconductor element at the tip of the probe. Electric current is passed through the conductor. In a magnetic field, a force is exerted on the moving electrons which tends to push them to one side of the conductor. A buildup of charge at the sides of the conductors will balance this magnetic influence, producing a measurable voltage between the two sides of the conductor. The presence of this measurable transverse voltage is called the Hall-effect after Edwin H. Hall, who discovered it in 1879.

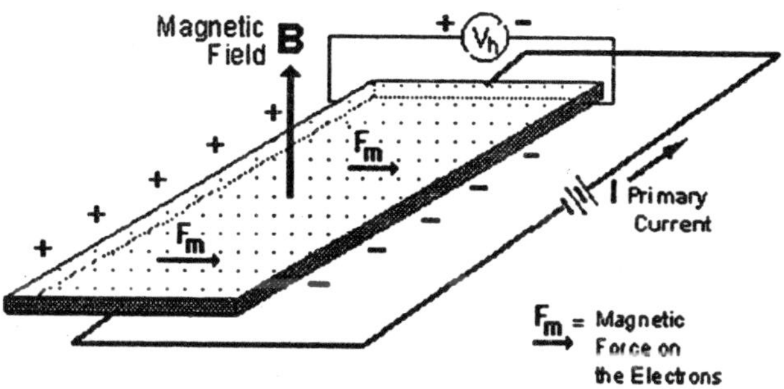

The voltage generated Vh can be related to the external magnetic field by the following equation.

Vh = I B Rh / b

Where:

Vh is the voltage generated.

I is the applied direct current.

B is the component of the magnetic field that is at a right angle to the direct current in the Hall element.

Rh is the Hall Coefficient of the Hall element.

b is the thickness of the Hall element.

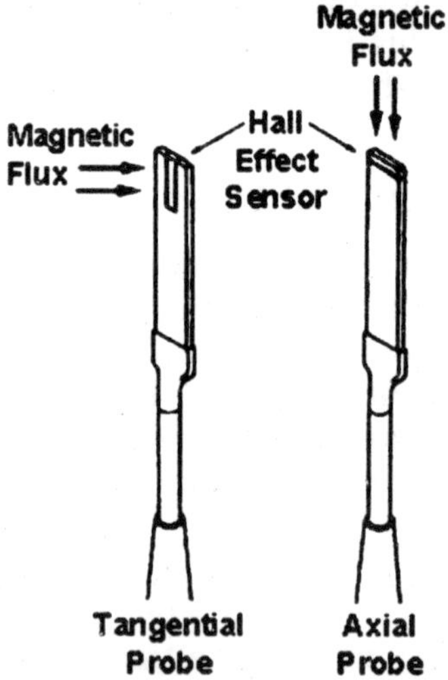

Probes are available with either tangential (transverse) or axial sensing elements. Probes can be purchased in a wide variety of sizes and configurations and with different measurement ranges. The probe is placed in the magnetic field such that the magnetic lines of force intersect the major dimensions of the sensing element at a right angle. Placement and orientation of the probe is very important and will be discussed in a later section.

PORTABLE MAGNETIZING EQUIPMENT FOR MAGNETIC PARTICLE INSPECTION

To properly inspect a part for cracks or other defects, it is important to become familiar with the different types of magnetic fields and the equipment used to generate them. As discussed previously, one of the primary requirements for detecting a defect in a ferromagnetic material is that the magnetic field induced in the part must intercept the defect at a 45 to 90 degree angle. Flaws that are normal (90 degrees) to the magnetic field will produce the strongest indications because they disrupt more of the magnet flux.

Magnetic Fields: A field of force that exists around a magnetic body or a current carrying conductor.

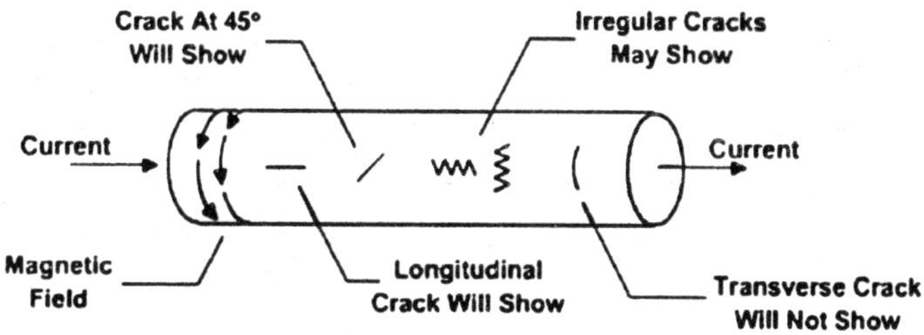

Therefore, for proper inspection of a component, it is important to be able to establish a magnetic field in at least two directions. A variety of equipment exists to establish the magnetic field for MPI. One way to classify equipment is based on its portability. Some equipment is designed to be portable so that inspections can be made in the field and some is designed to be stationary for ease of inspection in the laboratory or manufacturing facility. Portable equipment will be discussed first.

Permanent Magnets

Permanent magnets are sometimes used for magnetic particle inspection as the source of magnetism. The two primary types of permanent magnets are bar magnets and horseshoe (yoke) magnets. These industrial magnets are usually very strong and may require significant strength to remove them from a piece of metal. Some permanent magnets require over 50 pounds of force to remove them from the surface. Because it is difficult to remove the magnets from the component being inspected, and sometimes difficult and dangerous to place the magnets, their use is not particularly popular. However, permanent magnets are sometimes used by divers for inspection in underwater environments or other areas, such as explosive environments, where electromagnets cannot be used. Permanent magnets can also be made small enough to fit into tight areas where electromagnets might not fit.

Electromagnets

Today, most of the equipment used to create the magnetic field used in MPI is based on electromagnetism. That is, using an electrical current to produce the magnetic field. An electromagnetic yoke is a very common piece of equipment that is used to establish a magnetic field. It is basically made by wrapping an electrical coil around a piece of soft ferromagnetic steel. A switch is included in the electrical circuit so that the current and, therefore, the magnetic field can be turned on and off. They can be powered with alternating current from a wall socket or by direct current from a battery pack. This type of magnet generates a very strong magnetic field in a local area where the poles of the magnet touch the part being inspected. Some yokes can lift weights in excess of 40 pounds.

Prods

Prods are handheld electrodes that are pressed against the surface of the component being inspected to make contact for passing electrical current through the metal. The current passing between the prods creates a circular magnetic field around the prods that can be used in magnetic particle inspection. Prods are typically made from copper and have an insulated handle to help protect the operator. One of the prods has a trigger switch so that the current can be quickly and easily turned on and off. Sometimes the two prods are connected by any insulator (as shown in the image) to facilitate one hand operation. This is referred to as a dual prod and is commonly used for weld inspections.

If proper contact is not maintained between the prods and the component surface, electrical arcing can occur and cause damage to the component. For this reason, the use of prods are not allowed when inspecting aerospace and other critical components. To help prevent arcing, the prod tips should be inspected frequently to ensure that they are not oxidized, covered with scale or other contaminant, or damaged.

Portable Coils and Conductive Cables

Coils and conductive cables are used to establish a longitudinal magnetic field within a component. When a preformed coil is used, the component is placed against the inside surface on the coil. Coils typically have three or five turns of a copper cable within the molded frame. A foot switch is often used to energize the coil. Conductive cables are wrapped around the component. The cable used is typically 00 extra flexible or 0000 extra flexible. The number of wraps is determined by the magnetizing force needed and of course, the length of the cable. Normally, the wraps are kept as close together as possible. When using a coil or cable wrapped into a coil, amperage is usually expressed in ampere-turns. Ampere-turns is the amperage shown on the amp meter times the number of turns in the coil.

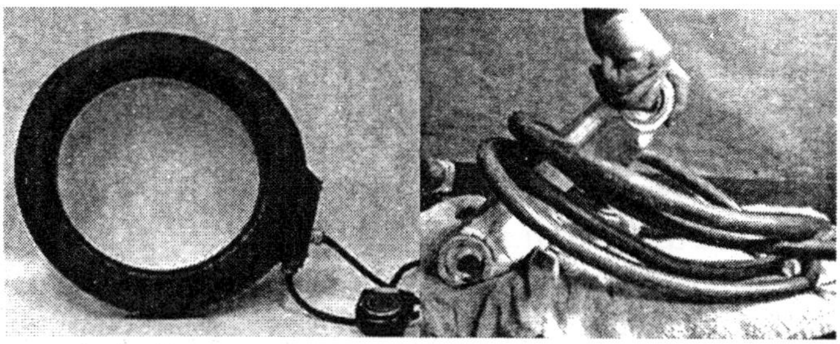

Fig. Portable Coil Conductive Cable

Portable Power Supplies

Portable power supplies are used to provide the necessary electricity to the prods, coils or cables. Power supplies are commercially available in a variety of sizes. Small power supplies generally provide up to 1,500A of half-wave direct current or alternating current when used with a 4.5 meter 0000 cable. They are small and light enough to be carried and operate on either 120V or 240V electrical service. When more power is necessary, mobile power supplies can be used. These units come with wheels so that they can be rolled where needed. These units also operate on 120V or 240V electrical service and can provide up to 6,000A of AC or half-wave DC when 9 meters or less of 0000 cable is used.

Stationary Equipment for Magnetic Particle Inspection

Stationary magnetic particle inspection equipment is designed for use in laboratory or production environment. The most common stationary system is the wet horizontal (bench) unit. Wet horizontal units are designed to allow for batch inspections of a variety of components. The units have head and tail stocks (similar to a lathe) with electrical contact that the part can be clamped between. A circular magnetic field is produced with direct magnetization. The tail stock can be moved and locked into place to accommodate parts of various lengths. To assist the operator in clamping the parts, the contact on the headstock can be moved pneumatically via a foot switch.

Most units also have a movable coil that can be moved into place so the indirect magnetization can be used to produce a longitudinal magnetic field. Most coils have five turns and can be obtained in a variety of sizes. The wet magnetic particle solution is collected and held in a tank. A pump and hose system is used to apply the particle solution to the components being inspected. Either the visible or fluorescent particles can be used. Some of the systems offer a variety of options in electrical current used for magnetizing the component. The operator has the option to use AC, half wave DC, or full wave DC. In some units, a demagnetization feature is built in, which uses the coil and decaying AC.

To inspect a part using a head-shot, the part is clamped between two electrical contact pads. The magnetic solution, called a bath, is then flowed over the surface of the part. The bath is then interrupted and a magnetizing current is applied to the part for a short duration, typically 0.5 to 1.5 seconds. (Precautions should be taken to prevent burning or overheating of the part.) A circular field flowing around the circumference of the part is created. Leakage fields from defects then attract the particles to form indications.

When the coil is used to establish a longitudinal magnetic field within the part, the part is placed on the inside surface of the coil. Just as done with a head shot, the bath is then flowed over the surface of the part. A magnetizing current is applied to the part for a short duration, typically 0.5 to 1.5 seconds, just after coverage with the bath is interrupted. Leakage fields from defects attract the particles to form visible indications.

The wet horizontal unit can also be used to establish a circular magnetic field using a central conductor. This type of a setup is used to inspect parts that have an open centre, such as gears, tubes, and other ring-shaped objects. A central conductor is an electrically conductive bar that is usually made of copper or aluminum. The bar is inserted through the opening and the bar is then clamped between the contact pads. When current is passed through the central conductor, a circular magnetic field flows around the bar and enters into the part or parts being inspected.

Multidirectional Equipment for Magnetic Particle Inspection

Multidirectional units allow the component to be magnetized in two directions, longitudinally and circumferentially, in rapid succession. Therefore, inspections are conducted without the need for a second shot. In multidirectional units, the two fields are balanced so that the field strengths are equal in both directions. These quickly changing balanced fields produce a multidirectional field in the component providing detection of defects lying in more than one direction.

Just as in conventional wet-horizontal systems, the electrical current used in multidirectional magnetization may be alternating, half-wave direct, or full-wave. It is also possible to use a combination of currents depending on the test applications. Multidirectional magnetization can be used for a large number of production applications, and high volume inspections.

To determine adequate field strength and balance of the rapidly changing fields, technique development requires a little more effort when multidirectional equipment is used. It is desirable to develop the technique using a component with known defects oriented in at least two directions, or a manufactured defect standard. Quantitative Quality Indicators (QQI) are also often used to verify the strength and direction of magnetic fields.

LIGHTS FOR MAGNETIC PARTICLE INSPECTION

Magnetic particle inspection can be performed using particles that are highly visible under white light conditions or particles that are highly visible under ultraviolet light conditions. When an inspection is being performed using the visible colour contrast particles, no special lighting is required as long as the area of inspection is well lit. A light intensity of at least 1000 lux (100 fc) is recommended when visible particles are used, but a variety of light sources can be used.

When fluorescent particles are used, special ultraviolet light must be used. Fluorescence is defined as the property of emitting radiation as a result of and during exposure to radiation. Particles used in fluorescent magnetic particle inspections are coated with a material that produces light in the visible spectrum when exposed to near-ultraviolet light. This "particle glow" provides high contrast indications on the component anywhere particles collect. Particles that fluoresce yellow-green are most common because this colour matches the peak sensitivity of the human eye under dark conditions. However, particles that fluoresce red, blue, yellow, and green colours are available.

Ultraviolet Light

Ultraviolet light or "black light" is light in the 1,000 to 4,000 Angstroms (100 to 400nm) wavelength range in the electromagnetic spectrum. It is a very energetic form of light that is invisible to the human eye. Wavelengths above 4,000A fall into the visible light spectrum and are seen as the colour violet. UV is separated according to wavelength into three classes: A, B, and C. The shorter the wavelength, the more energy that is carried in the light and the more dangerous it is to the human cells.

Class	Wavelength Range
UV-A	3,200–4,000 Angstroms
UV-B	2,800–3,200 Angstroms
UV-C	2,800–1,000 Angstroms

The desired wavelength range for use in nondestructive testing is between 3,500 and 3,800A with a peak wavelength at about 3,650A. This wavelength range is used because it is in the UV-A range, which is the safest to work with. UV-B will do an effective job of causing substances to fluoresce, however, it should not be used because harmful effects such as skin burns and eye damage can occur. This wavelength of radiation is found in the arc created during the welding process. UV-C (1,000 to 2,800A) is even more dangerous to living cells and is used to kill bacteria in industrial and medical settings.

The desired wavelength range for use in NDT is obtained by filtering the ultraviolet light generated by the light bulb. The output of a UV bulb spans a wide range of wavelengths. The short wavelengths of 3,120 to 3,340A are produced in low levels. A peak wavelength of 3650A is produced at a very high intensity. Wavelengths in the visible violet range (4050A to 4350A), green-yellow (5460A), yellow (6220A) and orange (6770A) are also usually produced. The filter allows only radiation in the range of 3200 to 4000A and a little visible dark purple to pass.

Basic Ultraviolet Lights

UV bulbs come in a variety of shapes and sizes. The more common types are the low pressure tube, high pressure spot, the high pressure flood types. The tubular black light is similar in construction to the tubular fluorescent lights used for office or home illumination. These lights use a low pressure mercury vapour arc. Tube lengths of 6 to 48 inches are common. The low pressure bulbs are most often used to provide general illumination to large areas rather than for illumination of components to be inspected. These bulbs generate a relatively large amount of white light, which is concerning since inspection specifications require less than two foot-candles of white light at the inspection surface.

Flood lights are also used to illuminate the inspection area, since they provide even illumination over a large area. Intensity levels for flood lamps are relatively low because the energy is spread over a large area. They generally do not generate the required UV light intensity at the given distance that specifications require.

Spot lights, on the other hand, provide concentrated energy that can be directed to the area of inspection. A spot light will generate a six inch diameter circle of high intensity light when held fifteen inches from the inspection surface. One hundred watt mercury vapour lights are most commonly used, but higher wattages are available.

In the high pressure mercury vapour spot or flood lamps, UV light is generated by a quartz tube inside the bulb. This tube contains two electrodes that establish an arc. The distance between electrodes is such that a starting electrode must be used. A resister limits the current to the starting electrode that establishes the initial arc that vapourizes the mercury in the tube. Once this low level arc is established and the mercury is vapourized, the arc between the main electrodes is established. It takes approximately five minutes to "warm up" and establish the arc between the main electrodes. This is why specifications require a "warm up time" before using the high pressure mercury vapour lights. Flood and spot black lights produce large amounts of heat and should be handled with caution to prevent burns. This condition has been eliminated by newer designs that include cooling fans. The arc in the bulb can be upset when exposed to an external magnetic field, such as that generated by a coil. Care should be taken not to bring the lamp close to strong magnetic fields, but if the arc is upset and extinguished, it must be allowed to cool before it can be safely restarted.

High Intensity Ultraviolet Lights

The 400 watt metal halide bulbs or "super lights" can be found in some facilities. This super bright light will provide adequate lighting over an area of up to ten times that covered by the 100 watt bulb. Due to their high intensity, excessive light reflecting from the surface of a component is a concern. Moving the light a greater distance from the inspection area will generally reduce this glare. Another type of high intensity light available is the micro-discharge light. This particular light produces up to ten times the amount of UV light conventional lights produce. Readings of up to 60,000 uW/cm^2 at 15 inches can be achieved.

MAGNETIC FIELD INDICATORS

Determining whether a magnetic field is of adequate strength and in the proper direction is critical when performing magnetic particle testing. As discussed previously, knowing the direction of the field is important because the field should be as close to perpendicular to the defect as possible and no more than 45 degrees from normal. Being able to evaluate the field direction and strength is especially important when inspecting with a multidirectional machine, because when the fields are not balanced properly, a vector field will be produced that may not detect some defects.

There is actually no easy-to-apply method that permits an exact measurement of field intensity at a given point within a material. In order to measure the field strength, it is necessary to intercept the flux lines. This is impossible without cutting into the material and cutting the material would immediately change the field within the part. However, cutting a small slot or hole into the material and measuring the leakage field that crosses the air gap with a Gauss meter is probably the best way to get an estimate of the actual field strength within a part. Nevertheless, there are a number of tools and methods available that are used to determine the presence and direction of the field surrounding a component.

Gauss Meter or Hall Effect Gage

A Gauss meter with a Hall Effect probe is commonly used to measure the tangential field strength on the surface of the part. As discussed in some detail on the "Measuring Magnetic Fields" page, the Hall effect is the transverse electric field created in a conductor when placed in a magnetic field. Gauss meters, also called Tesla meters, are used to measure the strength of a field tangential to the surface of the magnetized test object. The meters measure the intensity of the field in the air adjacent to the component when a magnetic field is applied.

The advantages of Hall effect devices are: they provide a quantitative measure of the strength of magnetizing force tangential to the surface of a test piece, they can be used for measurement of residual magnetic fields, and they can be used repetitively. Their main disadvantages are that they must be periodically calibrated and they cannot be used to establish the balance of fields in multidirectional applications.

Quantitative Quality Indicator (QQI)

The Quantitative Quality Indicator (QQI) or Artificial Flaw Standard is often the preferred method of assuring proper field direction and adequate field strength. The use of a QQI is also the only practical way of ensuring balanced field intensity and direction in multiple-direction magnetization equipment. QQIs are often used in conjunction with a Gauss meter to establish the inspection procedure for a particular component. They are used with the wet method only, and like other flux sharing devices, can only be used with continuous magnetization.

The QQI is a thin strip of either 0.002 or 0.004 inch thick AISI 1005 steel. A photoetch process is used to inscribe a specific pattern, such as concentric circles or a plus sign. QQIs are nominally 3/4 inch square, but miniature shims are also available. QQIs must be in intimate contact with the part being evaluated. This is accomplished by placing the shim on a part etched side down, and taping or gluing it to the surface. The component is then magnetized and particles applied. When the field strength is adequate, the particles will adhere over the engraved pattern and provide information about the field direction. When a multidirectional technique is used, a balance of the fields is noted when all areas of the QQI produce indications.

Some of the advantages of QQIs are: they can be quantified and related to other parameters, they can accommodate virtually any configuration with suitable selection, and they can be reused with careful application and removal practices. Some of the disadvantages are: the application process is somewhat slow, the parts must be clean and dry, shims cannot be used as a residual magnetism indicator as they are a flux sharing device, they can be easily damaged with improper handling, and they will corrode if not cleaned and properly stored.

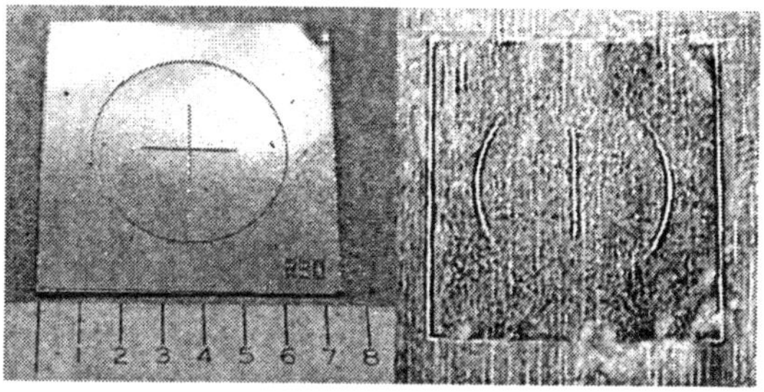

Above left is a photo of a typical QQI shim. The photo on the right shows the indication produced by the QQI when it is applied to the surface a part and a magnetic field is established that runs across the shim from right to left.

Pie Gage

The pie gage is a disk of highly permeable material divided into four, six, or eight sections by nonferromagnetic material. The divisions serve as artificial defects that radiate out in different directions from the centre. The diameter of the gage is 3/4 to 1 inch. The divisions between the low carbon steel pie sections are to be no greater than 1/32 inch. The sections are furnace brazed and copper plated. The gage is placed on the test piece copper side up and the test piece is magnetized. After particles are applied and the excess removed, the indications provide the inspector the orientation of the magnetic field.

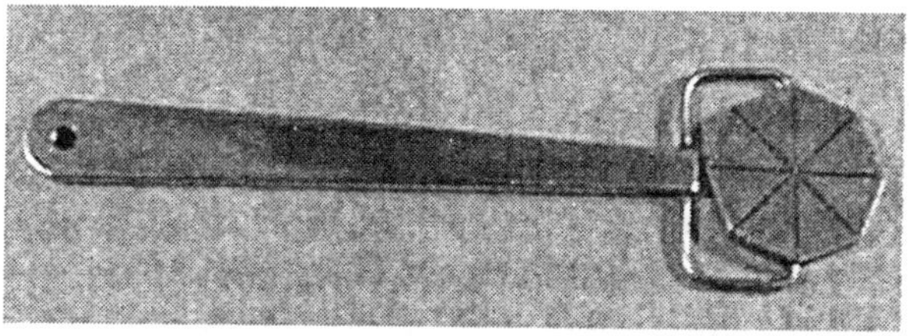

The principal application is on flat surfaces such as weldments or steel castings where dry powder is used with a yoke or prods. The pie gage is not recommended for precision parts with complex shapes, for wet-method applications, or for proving field magnitude. The gage should be demagnetized between readings.

Several of the main advantages of the pie gage are that it is easy to use and it can be used indefinitely without deterioration. The pie gage has several disadvantages, which include: it retains some residual magnetism so indications will prevail after removal of the source of magnetization, it can only be used in relatively flat areas, and it cannot be reliably used for determination of balanced fields in multidirectional magnetization.

Slotted Strips

Slotted strips, also known as Burmah-Castrol Strips, are pieces of highly permeable ferromagnetic material with slots of different widths. They are placed on the test object as it is inspected. The indications produced on the strips give the inspector a general idea of the field strength in a particular area.

Advantages of these strips are: they are relatively easily applied to the component, they can be used successfully with either the wet or dry method when using the continuous magnetization, they are repeatable as long as orientation to the magnetic field is maintained, and they can be used repetitively. Some of the disadvantages are that they cannot be bent to complex configuration and they are not suitable for multidirectional field applications since they indicate defects in only one direction.

Magnetic Particles

As mentioned previously, the particles that are used for magnetic particle inspection are a key ingredient as they form the indications that alert the inspector to defects. Particles start out as tiny milled (a machining process) pieces of iron or iron oxide. A pigment (somewhat like paint) is bonded to their surfaces to give the particles colour. The metal used for the particles has high magnetic permeability and low retentivity.

High magnetic permeability is important because it makes the particles attract easily to small magnetic leakage fields from discontinuities, such as flaws. Low retentivity is important because the particles themselves never become strongly magnetized so they do not stick to each other or the surface of the part. Particles are available in a dry mix or a wet solution.

Dry Magnetic Particles

Dry magnetic particles can typically be purchased in red, black, gray, yellow and several other colours so that a high level of contrast between the particles and the part being inspected can be achieved. The size of the magnetic particles is also very important. Dry magnetic particle products are produced to include a range of particle sizes. The fine particles are around 50 mm (0.002 inch) in size, and are about three times smaller in diameter and more than 20 times lighter than the coarse particles (150 mm or 0.006 inch). This make them more sensitive to the leakage fields from very small discontinuities. However, dry testing particles cannot be made exclusively of the fine particles. Coarser particles are needed to bridge large discontinuities and to reduce the powder's dusty nature. Additionally, small particles easily adhere to surface contamination, such as remnant dirt or moisture, and get trapped in surface roughness features. It should also be recognized that finer particles will be more easily blown away by the wind; therefore, windy conditions can reduce the sensitivity of an inspection. Also, reclaiming the dry particles is not recommended because the small particles are less likely to be recaptured and the "once used" mix will result in less sensitive inspections.

The particle shape is also important. Long, slender particles tend align themselves along the lines of magnetic force. However, research has shown that if dry powder consists only of long, slender particles, the application process would be less than desirable. Elongated particles come from the dispenser in clumps and lack the ability to flow freely and form the desired "cloud" of particles floating on the component. Therefore, globular particles are added that are shorter. The mix of globular and elongated particles result in a dry powder that flows well and maintains good sensitivity. Most dry particle mixes have particles with L/D ratios between one and two.

Wet Magnetic Particles

Magnetic particles are also supplied in a wet suspension such as water or oil. The wet magnetic particle testing method is generally more sensitive than the dry because the suspension provides the particles with more mobility and makes it possible for smaller particles to be used since dust and adherence to surface contamination is reduced or eliminated. The wet method also makes it easy to apply the particles uniformly to a relatively large area.

Wet method magnetic particles products differ from dry powder products in a number of ways. One way is that both visible and fluorescent particles are available. Most nonfluorescent particles are ferromagnetic iron oxides, which are either black or brown in colour. Fluorescent particles are coated with pigments that fluoresce when exposed to ultraviolet light. Particles that fluoresce green-yellow are most common to take advantage of the peak colour sensitivity of the eye but other fluorescent colours are also available. The particles used with the wet method are smaller in size than those used in the dry method for the reasons mentioned above. The particles are typically 10 mm (0.0004 inch) and smaller and the synthetic iron oxides have particle diameters around 0.1 mm (0.000004 inch). This very small size is a result of the process used to form the particles and is not particularly desirable, as the particles are almost too fine to settle out of suspension. However, due to their slight residual magnetism, the oxide particles are present mostly in clusters that settle out of suspension much faster than the individual particles. This makes it possible to see and measure the concentration of the particles for process control purposes. Wet particles are also a mix of long slender and globular particles.

The carrier solutions can be water or oil-based. Water-based carriers form quicker indications, are generally less expensive, present little or no fire hazard, give off no petrochemical fumes, and are easier to clean from the part. Water-based solutions are usually formulated with a corrosion inhibitor to offer some corrosion protection. However, oil-based carrier solutions offer superior corrosion and hydrogen embrittlement protection to those materials that are prone to attack by these mechanisms.

SUSPENSION LIQUIDS

Suspension liquids used in the wet magnetic particle inspection method can be either a well refined light petroleum distillate or water containing additives. Petroleum-based liquids are the most desirable carriers because they provided good wetting of the surface of metallic parts. However, water-based carriers are used more because of low cost, low fire hazard, and the ability to form indications quicker than solvent-based carriers. Water-based carriers must contain wetting agents to disrupt surface films of oil that may exist on the part and to aid in the dispersion of magnetic particles in the carrier. The wetting agents create foaming as the solution is moved about, so anti-foaming agents must be added. Also, since water promotes corrosion in ferrous materials, corrosion inhibitors are usually added as well.

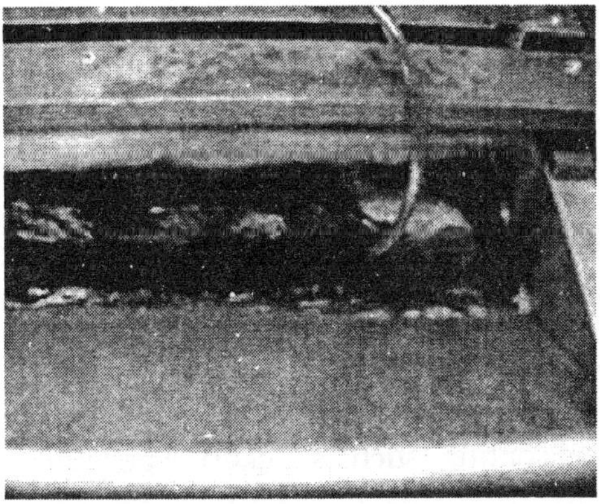

Petroleum based carriers are primarily used in systems where maintaining the proper particle concentration is a concern. The petroleum based carriers require less maintenance because they evapourate at a slower rate than the water-based carriers. Therefore, petroleum based carriers might be a better choice for a system that gets only occasional use or when regularly adjusting the carrier volume is undesirable. Modern solvent carriers are specifically designed with properties that have flash points above 200°F and keep nocuous vapours low. Petroleum carriers are required to meet certain specifications such as AMS 2641.

Dry Particle Inspection

In this magnetic particle testing technique, dry particles are dusted onto the surface of the test object as the item is magnetized. Dry particle inspection is well suited for the inspections conducted on rough surfaces. When an electromagnetic yoke is used, the AC or half wave DC current creates a pulsating magnetic field that provides mobility to the powder. The primary applications for dry powders are unground welds and rough as-cast surfaces.

Dry particle inspection is also used to detect shallow subsurface cracks. Dry particles with half wave DC is the best approach when inspecting for lack of root penetration in welds of thin materials. Half wave DC with prods and dry particles is commonly used when inspecting large castings for hot tears and cracks.

Steps in performing an inspection using dry particles

Prepare the part surface - the surface should be relatively clean but this is not as critical as it is with liquid penetrant inspection. The surface must be free of grease, oil or other moisture that could keep particles from moving freely. A thin layer of paint, rust or scale will reduce test sensitivity but can sometimes be left in place with adequate results. Specifications often allow up to 0.003 inch (0.076 mm) of a nonconductive coating (such as paint) and 0.001 inch max (0.025 mm) of a ferromagnetic coating (such as nickel) to be left on the surface. Any loose dirt, paint, rust or scale must be removed.

Apply the magnetizing force - Use permanent magnets, an electromagnetic yoke, prods, a coil or other means to establish the necessary magnetic flux.

Dust on the dry magnetic particles - Dust on a light layer of magnetic particles.

Gently blow off the excess powder - With the magnetizing force still applied, remove the excess powder from the surface with a few gentle puffs of dry air. The force of the air needs to be strong enough to remove the excess particles but not strong enough to dislodge particles held by a magnetic flux leakage field.

Terminate the magnetizing force - If the magnetic flux is being generated with an electromagnet or an electromagnetic field, the magnetizing force should be terminated. If permanent magnets are being used, they can be left in place.

Inspect for indications - Look for areas where the magnetic particles are clustered.

WET SUSPENSION INSPECTION

Wet suspension magnetic particle inspection, more commonly known as wet magnetic particle inspection, involves applying the particles while they are suspended in a liquid carrier. Wet magnetic particle inspection is most commonly performed using a stationary, wet, horizontal inspection unit but suspensions are also available in spray cans for use with an electromagnetic yoke. A wet inspection has several advantages over a dry inspection. First, all of the surfaces of the component can be quickly and easily covered with a relatively uniform layer of particles. Second, the liquid carrier provides mobility to the particles for an extended period of time, which allows enough particles to float to small leakage fields to form a visible indication. Therefore, wet inspection is considered best for detecting very small discontinuities on smooth surfaces. On rough surfaces, however, the particles (which are much smaller in wet suspensions) can settle in the surface valleys and lose mobility, rendering them less effective than dry powders under these conditions.

Steps in Performing an Inspection using wet Suspensions

Prepare the part surface - Just as is required with dry particle inspections, the surface should be relatively clean. The surface must be free of grease, oil and other moisture that could prevent the suspension from wetting the surface and preventing the particles from moving freely. A thin layer of paint, rust or scale will reduce test sensitivity, but can sometimes be left in place with adequate results. Specifications often allow up to 0.003 inch (0.076 mm) of a nonconductive coating (such as paint) and 0.001 inch max (0.025 mm) of a ferromagnetic coating (such as nickel) to be left on the surface. Any loose dirt, paint, rust or scale must be removed.

Apply the suspension - The suspension is gently sprayed or flowed over the surface of the part. Usually, the stream of suspension is diverted from the part just before the magnetizing field is applied.

Apply the magnetizing force - The magnetizing force should be applied immediately after applying the suspension of magnetic particles. When using a wet horizontal inspection unit, the current is applied in two or three short busts (1/2 second) which helps to improve particle mobility.

Inspect for indications - Look for areas where the magnetic particles are clustered. Surface discontinuities will produce a sharp indication. The indications from subsurface flaws will be less defined and lose definition as depth increases.

INSPECTION USING MAGNETIC RUBBER

The magnetic rubber technique was developed for detecting very fine cracks and is capable of revealing finer cracks than other magnetic techniques. Additionally, the technique can be use to examine difficult to reach areas, such as the threads on the inside diameter of holes, where the molded plugs can be removed and examined under ideal conditions and magnification if desired. The tradeoff, of course, is that inspection times are much longer.

The techniques uses a liquid (uncured) rubber containing suspended magnetic particles. The rubber compound is applied to the area to be inspected on a magnetized component. Inspections can be performed using either an applied magnetic field, which is maintained while the rubber sets (active field), or the residual field from magnetization of the component prior to pouring the compound. A dam of modeling clay is often used to contain the compound in the region of interest. The magnetic particles migrate to the leakage field caused by a discontinuity. As the rubber cures, discontinuity indications remain in place on the rubber.

The rubber is allowed to completely set, which takes from 10 to 30 minutes. The rubber cast is removed from the part. The rubber conforms to the surface contours and provides a reverse replica of the surface. The rubber cast is examined for evidence of discontinuities, which appear as dark lines on the surface of the molding. The molding can be retained as a permanent record of the inspection.

Magnetic rubber methods requires similar magnetizing systems used for dry method magnetic particle tests. The system may include yokes, prods, clamps, coils or central conductors. Alternating, direct current, or permanent magnets may be used to draw the particles to the leakage fields. The direct current yoke is the most common magnetization source for magnetic rubber inspection.

CONTINUOUS AND RESIDUAL MAGNETIZATION TECHNIQUES

In magnetic particle inspection, the magnetic particles can either be applied to the component while the magnetizing force is applied, or after it has been stopped. Continuous magnetization describes the technique where the magnetizing force is applied and maintained while the magnetic particles are dusted or flowed onto the surface of the component. In a wet horizontal testing unit, the application of the particles is stopped just before the magnetizing force is applied; but, since particles are still flowing over and covering the surface, this is considered continuous magnetization. Residual magnetization, on the other hand, describes the technique where the magnetizing force is applied to magnetize the component and then stopped before applying the magnetic particles. Only the residual field of the magnetized component is used to attract magnetic particles and produce an indication.

The continuous technique is generally chosen when maximum sensitivity is required because it has two distinct advantages over the residual technique. First, the magnetic flux will be highest when current is flowing and, therefore, leakage fields will also be strongest. Field strength in a component depends primarily on two variables: the applied magnetic field strength and the permeability of the test object. Viewing the upper right portion of the hysteresis loop, it is evident that the magnetic flux will be the strongest when the magnetizing force is applied. If the magnetizing force is strong enough, the flux density will reach the point of saturation. When the magnetizing force is removed, the flux density will drop to the retentivity point. The two gray traces show the paths the flux density would follow if the magnetizing force was applied and removed at levels that required to reach saturation. It can be seen that the flux density is always highest while the magnetizing current is applied. This is independent of the permeability of a material.

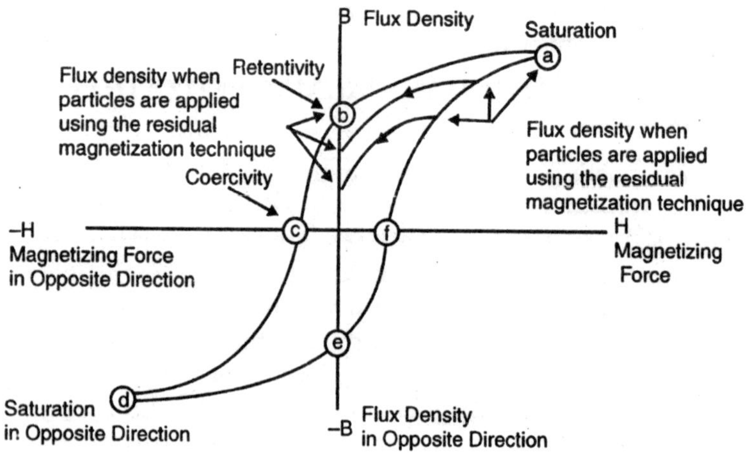

However, the permeability of the material is very important. High permeability materials do not retain a strong magnetic field so flux leakage fields will be extremely weak or nonexistent when the magnetizing force is removed. Therefore, materials with high magnetic permeability are not suited for inspection using the residual technique. When the residual technique is used to inspect materials with low permeability, care should be taken to ensure that the residual field is of the necessary strength to produce an indication. Defects should be relatively large and surface breaking to have a high probability of detection using the residual method.

The second advantage of the continuous technique is that when current is used to generate the magnetizing force, it can provide added particle mobility. Alternating or pulsed direct current will cause the particles to vibrate and move slightly on the surface of the part. This movement allows the particles to travel to leakage sites. More particles mean brighter indications compared to those formed using the residual technique.

One disadvantage of the continuous method is that heating of the component occurs when using direct magnetization. For example, when prods are used, they may create areas of localized heating when the continuous technique is used. This may be acceptable on components that will be further processed (removing this condition), but machined or in-service components may be adversely affected by this condition.

While generally not recommended, the residual technique does have its uses. It is commonly used in automated inspection systems to inspect materials with high retentivity. To speed throughput, automated systems often magnetize the parts and then submerge them in an agitated magnetic particle bath or pass them through a spray station. Closely controlled automated systems provided good results using the residual magnetism technique.

FIELD DIRECTION AND INTENSITY

Field Direction

As discussed previously, determining the direction of the field is important when conducting a magnetic particle inspection because the defect must produce a significant disturbance in the magnetic field to produce an indication. It is difficult to detect discontinuities that intersect the magnetic field at an angle less than 45°. When the orientation of a defect is not well established, components should be magnetized in a minimum of two directions at approximately right angles to each other. Depending on the geometry of the component, this may require longitudinal magnetization in two or more directions, multiple longitudinal and circular magnetization or circular magnetization in multiple directions. Determining strength and direction of the fields is especially critical when inspecting with a multidirectional machine. If the fields are not balanced, a vector field will be produced that may not detect some defects.

Depending on the application, pie gages, QQI's, or a gauss meter can be used to check the field direction. The pie gage is generally only used with dry powder inspections. QQI shims can be used in a variety of applications but are the only method recommended for use in establishing balanced fields when using multidirectional equipment.

Field Strength

The applied magnetic field must have sufficient strength to produce a satisfactory indication, but not so strong that it produces nonrelevant indications or limits particle mobility. If the magnetizing current is excessively high when performing a wet fluorescent particle inspection, particles can be attracted to the surface of the part and not be allowed to migrate to the flux leakage fields of defects. When performing a dry particle inspection, an excessive longitudinal magnetic field will cause furring. Furring is when magnetic particles build up at the magnetic poles of a part. When the field strength is excessive, the magnetic field is forced out of the part before reaching the end of the component and the poles along its length attract particles and cause high background levels. Adequate field strength may be determined by:

- Performing an inspection on a standard specimen that is similar to the test component and has known or artificial defects of the same type, size, and location as those expected in the test component. QQI shims can sometimes be used as the artificial defects.
- Using a gauss meter with a Hall effect probe to measure the peak values of the tangent field at the surface of the part in the region of interest. Most specifications call for a field strength of 30 to 60 gauss at the surface when the magnetizing force is applied.
- Looking for light furring at the ends of pipes or bars when performing dry particle inspections on these and other uncomplicated shapes.

Formulas for calculating current levels should only be used to estimate current requirements. The magnetic field strength resulting from calculations should be assessed for adequacy using one of the two method discussed above. Likewise, published current level information should also be used only as a guide unless the values have been established for the specific component and target defects of the inspection at hand.

Using a Pie Gage

A pie gage is placed copper side up and held in contact with the component as the magnetic field and particles are applied. Indications of the leakage fields provide a visual representation of defect direction within the component. Pie gages work well on flat surfaces, but if the surface is concave or convex, inaccurate readings may occur. The pie gage is a flux sharing device and requires good contact to provide accurate readings.

Using Quantitative Quality Indicator (QQI) Shims

Quantitative Quality Indicator (QQI) flaw shims are used to establish proper field direction and to ensure adequate field strength during technique development. The QQI flaw shim is the most efficient means of determining balance and effectiveness of fields. The QQI's are also flux sharing devices and must be properly attach so as not to allow particles to become trapped under the artificial flaw. Application using Super glue is the preferred way of attaching the artificial flaw, but does not allow for reuse of the shims. Shims can also be attached with tape applied to just the edge of the shim. It is recommended that the tape be impervious to oil, not be fluorescent, and be 1/4 to 1/2 inch in width.

The QQI must be applied to locations on the component where the flux density may vary. One example would be the centre area of a yoke or Y shaped component. Oftentimes, the flux density will be near zero in this area. If two legs of a Y are in contact with the pad in circular magnetization, it must be determined if current is flowing evenly through each leg. A QQI on each leg would be appropriate under such conditions.

QQI's can be used to establish system threshold values for a defect of a given size. By attaching a QQI shim with three circles (40percent, 30percent and 20percent of shim thickness) to the component, threshold values for a specific area of the component, can be established. Begin by applying current at a low amperage and slowly increasing it until the largest flaw is obtained. The flux density should be verified and recorded using a Hall effects probe. The current is then increased until the second circle is identified and the flux density is again recorded. As the current is further increased, the third ring is identified and the current values are recorded.

Hall Effects Gauss Meter

There are several types of Hall effects probes that can be used to measure the magnetic field strength. Transverse probes are the type most commonly used to evaluate the field strength in magnetic particle testing. Transverse probes have the Hall effect element mounted in a thin, flat stem and they are used to make measurements between two magnetic poles. Axial probes have the sensing element mounted such that the magnetic flux in the direction of the long axis of the probe is measured.

To make a measurement with a transverse probe, the probe is positioned such that the flat surface of the Hall effect element is transverse to the magnetic lines of flux. The Hall effect voltage is a function of the angle at which the magnetic lines of flux pass through the sensing element. The greatest Hall effect voltage occurs when the lines of flux pass perpendicularly through the sensing element. If not perpendicular, the output voltage is related to the cosine of the difference between 90 degrees and the actual angle. The peak field strength should be measured when the magnetizing force is applied. The field strength should be measured in all areas of the component to be inspected.

PARTICLE CONCENTRATION AND CONDITION

Particle Concentration

The concentration of particles in the suspension is a very important parameter in the inspection process and must be closely controlled. The particle concentration is checked after the suspension is prepared and regularly monitored as part of the quality system checks. ASTM E-1444-01 requires concentration checks to be performed every eight hours or at ever shift change.

The standard process used to perform the check requires agitating the carrier for a minimum of thirty minutes to ensure even particle distribution. A sample is then taken in a pear-shaped 100 ml centrifuge tube having a stem graduated to 1.0 ml in 0.05 ml increments for fluorescent particles, and graduated to 1.5 ml. in 0.1 ml increments for visible particles. The sample is then demagnetized so that the particles do not clump together while settling. The sample must then remain undisturbed for a minimum of 60 minutes for a petroleum-based carrier or 30 minutes for a water-based carrier, unless shorter times have been documented to produce results similar to the longer settling times. The volume of settled particles is then read. Acceptable ranges are 0.1 to 0.4 ml for fluorescent particles and 1.2 to 2.4 ml for visible particles. If the particle concentration is out of the acceptable range, particles or the carrier must be added to bring the solution back in compliance with the requirement.

Particle loss is often attributed to "dragout." Dragout occurs because the solvent easily runs off components and is recaptured in the holding tank. Particles, on the other hand, tend to adhere to components, or be trapped in geometric features of the component. These particles will be "drug out" or lost to the system and will eventually need to be replaced.

Particle Condition

After the particles have settled, they should be examined for brightness and agglomeration. Fluorescent particles should be evaluated under ultraviolet light and visible particles under white light. The brightness of the particles should be evaluated weekly by comparing the particles in the test solution to those in an unused reference solution that was saved when the solution was first prepared. The brightness of the two solutions should be relatively the same. Additionally, the particles should appear loose and not lumped together. If the brightness or the agglomeration of the particles is noticeably different from the reference solution, the bath should be replaced.

SUSPENSION CONTAMINATION

The suspension solution should also be examined for evidence of contamination. Contamination primarily comes from inspected components. Oils, greases, sand, and dirt will be introduced to the system through components. If the area is unusually dusty, the system will pickup dust or other contaminates from the environment.

This examination is performed on the carrier and particles collected for concentration testing. The graduated portion of the tube is viewed under ultraviolet and white light when fluorescent particles are being used, and under white light when visible particles are being used. The magnetic particles should be examined for foreign particles, such as dirt, paint chips and other solids. Differences in colour, layering or banding within the settled particles would indicate contamination. Some contamination is to be expected but if the foreign matter exceeds 30 percent of the settled solids, the solution should be replaced.

The liquid carrier portion of the solution should also be inspected for contamination. Oil in a water bath and water in a solvent bath are the primary concerns. If the solution fluoresces brightly when fluorescent particles are being used, this can be an indication that dye is being dislodged from the particles by the mixing pump. While not technically contamination, this condition should be further evaluated by allowing the collected sample bath to set for 10 to 12 hours and viewed under ultraviolet light. If a band that fluoresces brighter than the bulk of particles is evident on top of the settled solids, the bath contains excessive unattached fluorescent pigments and should be discarded.

Water Break Test

A daily water break check is required to evaluate the surface wetting performance of water-based carriers. The water break check simply involves flooding a clean surface similar to those being inspected and observing the surface film. If a continuous film forms over the entire surface, sufficient wetting agent is present. If the film of suspension breaks (water break) exposing the surface of the component, insufficient wetting agent is present and the solution should be adjusted or replaced.

ELECTRICAL SYSTEM CHECKS

Changes in the performance of the electrical system of a magnetic particle inspection unit can obviously have an effect on the sensitivity of an inspection. Therefore, the electrical system must be checked when the equipment is new, when a malfunction is suspected, or every six months.

Maximum Unit Output

A magnetic particle unit should be checked regularly (daily) to verify that the maximum current output of its electrical system has not changed over time. This check is performed by placing a large diameter (typically 1 inch or more) copper or aluminum bar between the contacts and energizing the circuit with the current control set to the maximum value. The maximum amperage as recorded by the system's ammeter should be noted and compared to the value obtained when the unit was last certified or calibrated. A significant drop or gradual decay indicates a problem with the unit that should be corrected.

Internal Short Test

The internal short circuit testing is required to determine the if current is "leaking past" the test article. Internal electrical leakage will cause the test equipment to indicate current is flowing through the component when it is actually bypassing the component. The process for checking equipment for this condition is to energize the system with nothing between the contact heads and the amperage set to the highest output. If the ammeter shows any current flow, internal shorting should be suspected.

Ammeter Check

It is important that the ammeter provide consistent and correct readings. If the meter is reading low, over magnetization will occur and possibly result in excessive background fluorescence. If ammeter readings are high, flux density could be too low to produce detectable indications. To verify ammeter accuracy, a calibrated ammeter is connected in series with the output circuit and values are compared to the equipment's ammeter values. Readings are taken at three output levels in the working range. The equipment meter is not to deviate from the calibrated ammeter more than +/- 10 percent or 50 amperes, whichever is greater. If the meter is found to be outside this range, the condition must be corrected. It should be noted that when measuring half-wave rectified current, the readings should be doubled.

Quick Break Test

A quick break circuit is primarily used in three-phase, full-wave rectified systems to ensure longitudinal magnetic fields maintain their strength to the end of a component. Quick break circuits cause the current to abruptly collapse or drop to zero at the end of the activation cycle. By doing this, low frequency eddy currents are produced close to the surface of a component. These eddy currents help extend the useable longitudinal field to the ends of the component. To determine if a quick break system is functioning properly, an oscilloscope is often used. By observing the sine wave of the current, it can be determined if current drops abruptly when the circuit is interrupted.

Shot Timer Check

When a timer is used to control the shot duration, the timer must be calibrated. ASTM E-1444 -01 requires the timer be calibrated to within +/- 0.1 second. A certified timer should be used to verify the equipment timer is within the required tolerances.

LIGHTING

Magnetic particle inspection predominately relies on visual inspection to detect any indications that form. Therefore, lighting is a very important element of the inspection process. Obviously, the lighting requirements are different for an inspection conducted using visible particles than they are for an inspection conducted using fluorescent particles.

Light Requirements When Using Visible Particles

Magnetic particle inspections that use visible particles can be conducted using natural or artificial lighting. When using natural lighting, it is important to keep in mind that daylight varies from hour to hour. Inspector must constantly stay aware of the lighting conditions and make adjustments when needed. To improve the uniformity of lighting from one inspection to the next, the use of artificial lighting is recommended. Artificial lighting should be white whenever possible and white flood or halogen lamps are most commonly used. The light intensity is required to be 100 foot-candles at the surface being inspected. It is advisable to choose a white light wattage that will provide sufficient light, but avoid excessive reflected light that could distract from the inspection.

Light Requirements When Using Fluorescent Particles

Ultraviolet Lighting

When performing a magnetic particle inspection using fluorescent particles, the condition of the ultraviolet light and the ambient white light must be monitored. Standards and procedures require verification of lens condition and light intensity. Black lights should never be used with a cracked filter as the output of white light and harmful black light will be increased. The cleanliness of the filter should also be checked regularly since a coating of solvent carrier, oil, or other foreign material can reduce the intensity or light by as much as 50percent. The filter should be checked visually and cleaned as necessary before warming-up the light.

For UV lights used in component evaluations, the normally accepted intensity is 1000mwatts/cm^2 when measured at 15 inches from the filter face (requirements can vary from 800 to 1200mwatts/cm^2). The required check should be performed when a new bulb is installed, at startup of the inspection cycle, if a change in intensity is noticed, or every eight hours if in continuous use. Regularly checking the intensity of UV lights is very important because bulbs lose intensity over time. In fact, a bulb that is near the end of its operating life will often have an intensity of only 25percent of its original output. Black light intensity will also be affected by voltage variations, so it is important to provide constant voltage to the light. A bulb that produces acceptable intensity at 120 volts will produce significantly less at 110 volts.

Ambient White Lighting

When performing a fluorescent magnetic particle inspection, it is important to keep white light to a minimum as it will significantly reduce the inspectors ability to detect fluorescent indications. Light levels of less than 2fc are required by most procedures and some procedures require it to be less than 0.5fc at the inspection surface. Some specifications require that a white light intensity measurement be made at 15 inches from a UV light source to verify that the white light is being removed by the filter.

White Light for Indication Confirmation

While white light is held to a minimum in fluorescent inspections, procedures may require that indications be evaluated under white light. The white light requirements for this evaluation are the same as when performing an inspection with visible particles. The minimum light intensity at the surface being inspected must be 100fc.

Light Measurement

Light intensity measurements are made using a radiometer. A radiometer is an instrument that translates light energy into an electrical current. Light striking a silicon photodiode detector causes a charge to build up between internal layers. When an external circuit is connected to the cell, an electrical current is produced. This current is linear with respect to incident light. Some radiometers have the ability to measure both white and UV light, while others require a separate sensor for each measurement. The sensing area should always be kept clean and free of materials that could reduce or obstruct light reaching the sensor.

Radiometers are relatively unstable instruments and readings often change considerably over time. Therefore, they must be calibrated regularly. They should be calibrated at least every six months. A unit should be checked to make sure its calibration is current before taking any light readings.

Ultraviolet light measurements should be taken using a fixture to maintain a minimum distance of 15 inches from the filter face to the sensor. The sensor should be centred in the light field to obtain the peak reading. UV spot lights are often focused which causes intensity readings to vary considerable over a small area. White lights are seldom focused and, depending on the wattage, will often produce in excess of the 100fc at 15 inches.

EYE CONSIDERATION

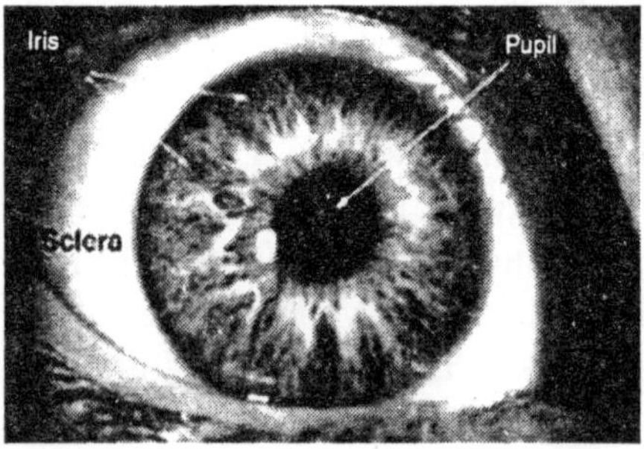

Eye Adaptation

Just as lighting is an important consideration in the inspection process, so is the eye's response to light. Scientists have recently discovered that a special, tiny group of cells at the back of the eye help tell the brain how much light there is, causing the pupil to get bigger or smaller. The change in pupil diameter is not instantaneous, therefore, eyes must be given time to adapt to changing lighting conditions. When performing a fluorescent magnetic particle inspection, the eye must be given time to adapt to the darkness of the inspection booth before beginning to look for indications. Dark adaptation time of at least one minute is required by most procedures. Some studies recommend adaptation time of five minutes if entering an inspection area from direct sunlight. Inspectors should carefully adhere to the required adaptation time as it is quite easy to overlook an indication when an inspection is started before the eyes have adjusted to the darkened conditions.

Eyeball Fluorescence

When ultraviolet light enters the human eye, the fluid that fills the eye fluoresces. This condition is called eyeball fluorescence, and while it is considered harmless, it is annoying and interferes with vision while it exists. When working around ultraviolet lights, one should be careful not to look directly into lights and to hold spot lights to avoid reflection. UV light will be reflected from surfaces just as white light will, so it is advisable to consider placement of lights to avoid this condition. Special filtered glasses may be worn by the inspector to remove all UV light from reaching the eyes but allowing yellow-green light from fluorescent indications to pass. Technicians should never wear darkened or photochromatic glasses as these glasses also filter or block light from fluorescent indications.

Examples of Visible Dry Magnetic Particle Indications

Eddy Current: A current induced in a conductor situated in a changing magnetic field or moving in a fixed one.

Ultrasonic: The study and use of pressure waves that have a frequency in excess of 20,000 Hz and therefore inaudible to the human ear.

One of the advantages that a magnetic particle inspection has over some of the other nondestructive evaluation methods is that flaw indications generally resemble the actual flaw. This is not the case with NDT methods such as ultrasonic and eddy current inspection, where an electronic signal must be interpreted. When magnetic particle inspection is used, cracks on the surface of the part appear as sharp lines that follow the path of the crack. Flaws that exist below the surface of the part are less defined and more difficult to detect. Below are some examples of magnetic particle indications produced using dry particles.

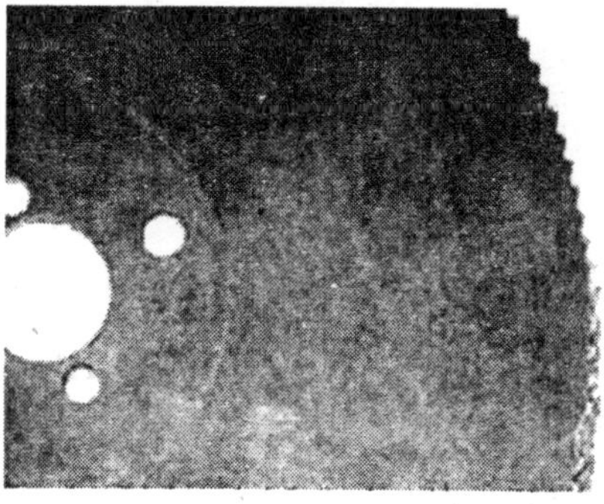

Fig. Indication of a crack in a saw blade

Fig. Indication of cracks in a weldment

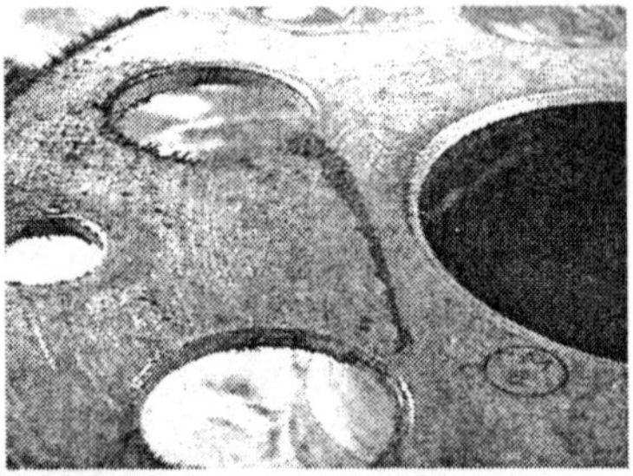

Fig. Indication of cracks originating at a fastener hole

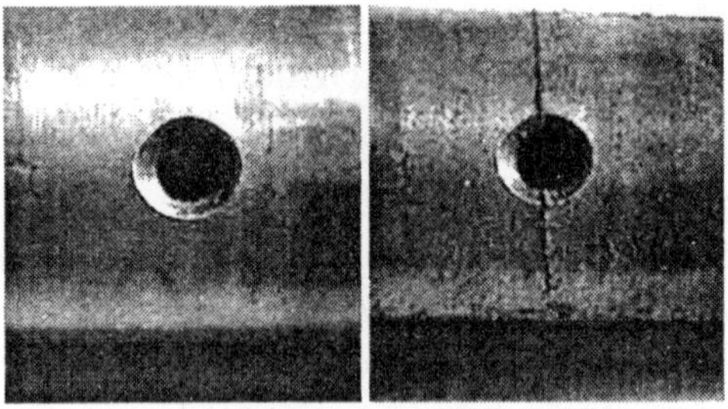

Fig. Before and after inspection pictures of cracks emanating from a hole

Fig. Indication of cracks running between attachment holes in a hinge

EXAMPLES OF FLUORESCENT WET MAGNETIC PARTICLE INDICATIONS

The indications produced using the wet magnetic particles are more sharp than dry particle indications formed on similar defects. When fluorescent particles are used, the visibility of the indications is greatly improved because the eye is drawn to the "glowing" regions in the dark setting. A few examples of fluorescent wet magnetic particle indications.

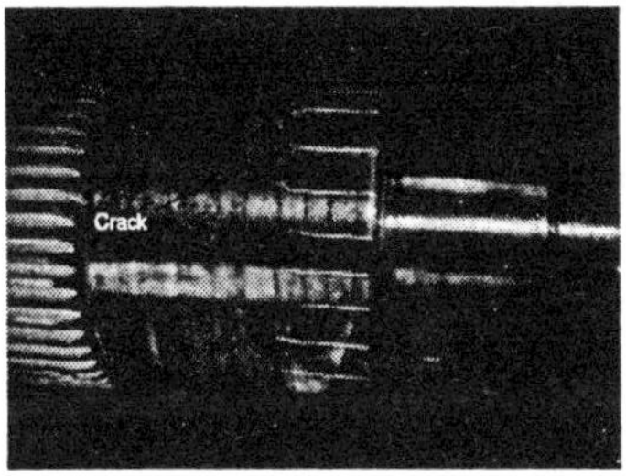

Fig. Magnetic particle wet fluorescent indication of cracks in a drive shaft.

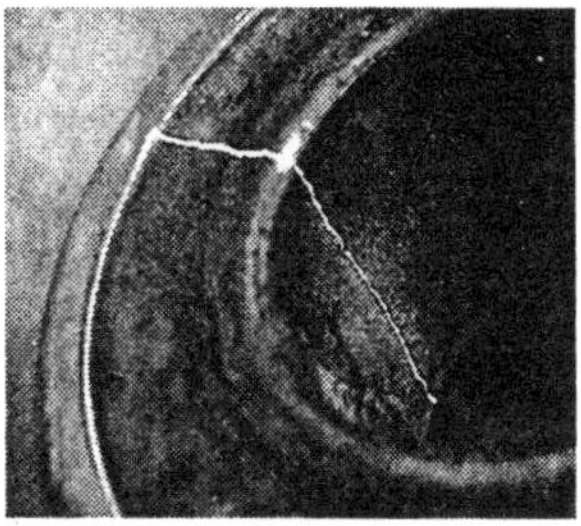

Fig. Magnetic particle wet fluorescent indication of a crack in a bearing.

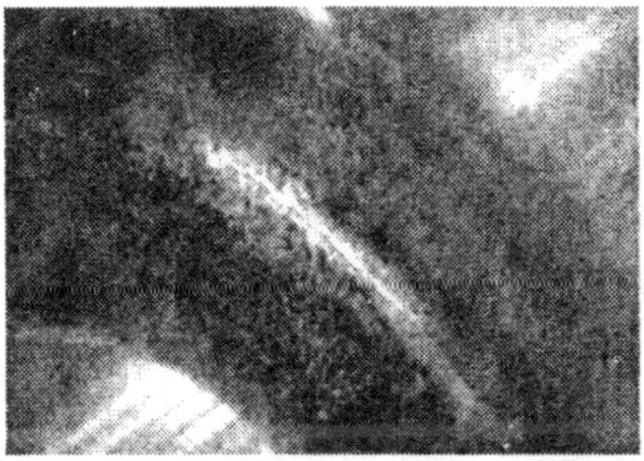

Fig. Magnetic particle wet fluorescent indication of a crack in a crane hook.

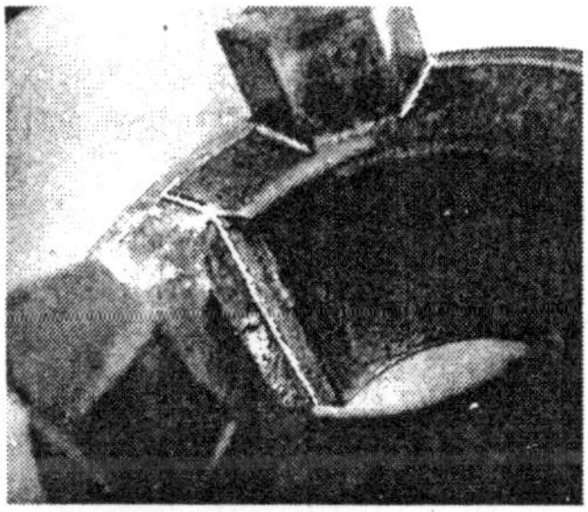

Fig. Magnetic particle wet fluorescent indication of a crack at a sharpradius.

Fig. Magnetic particle wet of fluorescent indication a crack in a casting.

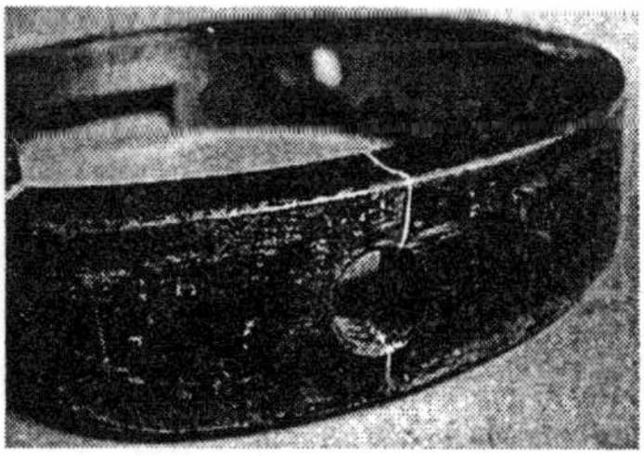

Fig. Magnetic particle wet fluorescent indication of cracks at a fastener hole.

Chapter 4

Field Theory from Electromagnetism

ELECTROMAGNETIC FIELD THEORY

It was the scientist Michael Faraday who first studied in detail the phenomena involving the interaction between electricity and magnetism. Amongst his many achievements, he is credited with the construction of the first electric motor and the discovery of both the principle and the method whereby a rotating magnet can be used to create an electric current in a coil of wire.He also observed the way in which iron filings arrange around a magnet. They appear to follow lines of magnetism leading out from one magnetic pole and back to the other pole. He suggested that the effect of a magnet on a wire carrying a current is the result of "lines of force". Faraday spent the latter part of his life working to devise an experiment that would confirm his theory of electromagnetic fields but without success.

Finally, a Scottish mathematician, James Clerk Maxwell, picked up on Faraday's ideas and was successful in developing a theory of electromagnetism which enabled testable predictions to be made. Inspired by Faraday's lines of force, he developed a model that unified magnetic and electrical forces. Maxwell "unified" the study of electricity and magnetism in four tidy equations. In essence he discovered that electric and magnetic fields were intrinsically related to one another, with or without the presence of a conductive path for electrons to flow. Stated simply Maxwell's discovery was this:-

'A changing electric field produces a perpendicular magnetic field' and 'A changing magnetic field produces a perpendicular electric field'

Quite interestingly Maxwell predicted that electromagnetism would be propagated through space at a finite rate and was struck by the similarity between the predicted speed of electromagnetism and the speed of light. From this connection sprang the idea that light was an electric phenomenon and the subsequent discovery of radio waves.

Classical electromagnetic theory has been remarkably robust to modern innovations. It was the inspiration for special relativity and did not need alteration. As quantum mechanics evolved, the field equations remained the same. One technical change was in the action, a gauge had to be chosen. The advent of general relativity did not alter electromagnetism. Instead, general relativity provided the background metric needed by electromagnetism. The U(1) symmetry of electromagnetism served as the starting point for the formation of the standard model, which unifies electromagnetism, the weak, and the strong forces. A problem with this picture is the separation of gravity from the forces of the standard model. In this paper, I use electromagnetic theory to deduce the laws that govern gravity.

THE ACTION

The classical action for electromagnetism has three terms, a contraction of the irreducible anti symmetric field strength tensor and a coupling of electric charge with the potential, and an inertia term. To be logically complete, Nature must also exploit a contraction of an irreducible symmetric field strength tensor. The symmetric field will be associated with a different charge current that couples with the potential because the tensor is irreducible and thus an independent fundamental force. This new force must be long range like electromagnetism. The only such force is gravity.

Like charges repel in the electromagnetic action for several reasons. First, the signs of the charge coupling and the field strength tensor contraction are the same. The field equations that arise from varying the action with respect to the potential keeping the velocity fixed will have like charges repel. Second, the signs of the charge coupling term and the inertia term are the same. Varying the action with respect to velocity keeping the potential fixed generates the Lorentz force where like charges repel. Third, the field strength tensor is anti symmetric, so it must be represented by an odd spin particle, the spin 1 photon. For gravity, like charges attract. The coupling term must have the sign opposite to the field strength contraction and inertia terms so both the field and force equations will have like charges attract. A traceless symmetric rank 2 tensor can be represented by a spin 2 particle that travels at the speed of light like the photon, so the particle matches the description of a graviton. Based on these deductions, the action for gravity and electromagnetism (GEM) can be written:

$$S_{GEM} = \int \sqrt{-g}\, d^4x \Big(-\frac{\rho_m}{\gamma} - \frac{1}{c} J^\mu (A_\mu + I_{\|} A_\mu I_{\|}) + \frac{1}{c} J^\mu (I_{tr}(A_\mu - I_{\|} A_\mu i_{\|}) I_{tr})^* \tag{1}$$

$$-\frac{1}{4c^2}(\nabla_\mu A_\nu - \nabla_\nu A_\mu)(\nabla^\mu A^\nu - \nabla^\nu A^\mu) - \frac{1}{4c^2}(\nabla_\mu A_\nu + \nabla_\nu A_\mu)(\nabla^\mu A^\nu + \nabla^\nu A^\mu)\Big).$$

The current coupling term appears more complex than one might anticipate. The reason is that this current coupling must represent both a spin 1 and a spin 2 current-current interaction. One is free to choose the orientation of the theoretical 4-potential with the actual 4-current, derived from the 4-potential via a Fourier transformation. Where there is freedom, there is a physics message. In this case, we have freedom with the phase for the current-current interaction. The part of the potential that is tranverse to the current can be separated from that which is parallel. The phase of the current-current interaction for the transverse potential-current has the symmetry of a spin 1 field, as is needed to have like charges repel. Rotating the longitudinal current around the parallel axis accomplishes two things. First, the sign of the coupling term must change to conserver the correct signs of the contraction of the current density with the potential. Second, the phase of the current-current interaction for the rotated longitudinal potential-current has the symmetry of a spin 2 field. A even spin field is required for like charges to attract.

The current density is made of difference between the electric current density and the mass current density, using the square root of Newton's gravitational constant with the mass current density so the units are identical. Electric charge is only known to ten significant digits, while for fundamental particles, the mass in electrical units can be more than thirteen orders of magnitude smaller. At this time, I have no notion why there is this large difference, but it justifies the practice of working on problems in electromagnetism separately from gravity.

The deep insight into gravity provided by general relativity is that the geometry of spacetime is not presumed. In special relativity, the metric must be constant, the same for all inertial observers. In general relativity, the metric can vary depending on where the observer is in spacetime: closer to a mass source will have more spacetime curvature. The Riemann curvature tensor is a rank 4 tensor that contains second order derivatives of the metric. The Hilbert action has the Ricci scalar, a contraction of the Riemann curvature tensor. By varying the Hilbert action with respect to the metric tensor, the geometry of spacetime can change as dictated by the Einstein field equations.

Electromagnetism works no matter what the metric happens to be. The antisymetric tensor cannot characterize how the symmetric metric tensor changes, so the possibly dynamic metric must be supplied as part of the background mathematical structure. By including a symmetric field strength tensor in the GEM action, it becomes possible for the action to account for the changes in metric tensor, and thus remove the metric from the background structure. The idea is not to treat the metric as an active field, but instead to use a symmetry of the action to provide a constraint on a metric that varies in spacetime.

The way gravity in the GEM action works mathematically is different from general relativity. The Hilbert action requires only one field, the metric field, to be varied in the action to generate the Einstein field equations. The GEM action does not treat the metric as a field. Instead there is a diffeomorphism symmetry of the action that allows one to chose how much of a covariant derivative is due to a dynamic potential or a dynamic metric. Where there is a symmetry, there is a conserved charge. Since the symmetry involves the metric, the charge must be mass. Energy and momentum also arise from a symmetry of the action. Since mass is the square of the energy minus the square of the momentum, it is logically consistent that energy, momentum, and mass all can be viewed as different symmetries of the same action.

The action has symmetry for mass less particles. Mass charge breaks symmetry. This is an attractive feature because it means the Higgs mechanism for introducing mass into the standard model is unnecessary. The GEM proposal predicts the Higgs boson will not be found in particle field experiments such as the Large Hadron Collider. The trace of the symmetric field strength tensor is a scalar field which plays the role of the Higgs.

FIELD EQUATIONS AND SOLUTIONS

The field equations can be determined in the usual way by fixing the metric tensor up to a diffeomorphism and then varying the action with respect to the potential:

$$J^{\mu}_{tr} - J^{\mu}_{\|} = \nabla_{\mu}\nabla^{\mu}A^{\mu}$$

This 4D wave equation has the ability to describe 4 modes of emission of particles that travel at the speed of light. The two transverse modes are electromagnetism. That leaves the scalar and longitudinal modes for the graviton.

It is critical to note in equation. wave the scalar D'Alembertian operator is not acting alone on the potential. Instead, two covariant derivatives are applied to the potential. A covariant derivative contains both the normal derivative and the connection. The complete wave equation is composed of the D'Alembertian operator acting on the potential, the divergence of the connection, the connection of the first derivative of the potential, and the connection of the connection. Spacetime is nearly flat if one excludes regions near black holes, so the connection of the connection term can be ignored. The divergence of the connection contains second derivatives of the metric, so in principle could provide an equation that must be solved to determine a smoothly changing metric.

Because equation wave is valid up to a diffeomorphism of the metric, the equation cannot be solved until that degree of freedom is eliminated. In this paper, I will work with metric compatible, torsion-free connections as is done in general relativity. If one chooses to work with a flat Minkowski metric in Euclidean spacetime, the connection will be zero everywhere. For a static point charge, the first field equation is a sum of the static form of Gauss' law of electric charge and Newton's law of gravity:

$$\rho_{Ir} - \rho_{||} = -c\nabla^2\phi$$

The repulsive force of like electric charges is decreased by mass attraction. This field equation can account for all classical effects of Gauss' and Newton's law because it is identical to their superposition. There are two articles in the literature that assert without proof that one cannot form vector field equations where like charges attract. Perhaps there was an aversion to putting a negative sign in front of the mass. What matters are the relative signs, that the mass charge density has the same sign as the Laplacian operator in equation.

A flaw of Newton's law of gravity is that it requires changes in the charge density to propagate instantaneously, which is not consistent with special relativity. One road to Einstein's field equations is to make Newton's law respect special relativity. Here is the dynamic form of the equation

$$\rho_{tr} - \rho_{\parallel} = \frac{1}{c}\frac{\partial^2 \phi}{\partial t^2} - c\nabla^2\phi$$

The waves for gravity and light both propagate at the speed of light. This equation transforms as the first term of a 4-vector equation and so is manifestly covariant under a Lorentz transformation.

The gauge choice - how one measures things - for the GEM action concerns how to measure a covariant derivative: how much should be due to the standard derivative versus how much is due to the connection, the changes in the metric? In the preceding paragraphs, a choice was made that everything was due to the standard derivative of the potential. This time I choose to allow all of the change be due to the connection, nothing due to the standard derivative of the potential. The first field equation under this gauge choice for a static charge is:

$$\rho_{tr} - \rho_{\parallel} = -\partial_{\mu}\Gamma^{0\mu}_{\mu} A^{\mu}$$

A metric must be found whose connection solves this first-order partial differential equation. This might appear like a difficult task since the connection has three derivatives of the metric. Yet one of the derivatives is zero because the charges are static. If the metric were diagonal, another of the derivatives would be zero. The metric must reduce to the Minkowski metric as the charges go to zero. The exponential metric below provides a solution with these properties to equation ChristoffelDivergence:

$$g_{\mu\nu} = \begin{pmatrix} exp(2(\frac{\sqrt{G}q - GM}{c^2 R})) & 0 & 0 & 0 \\ 0 & -exp(-2(\frac{\sqrt{G}q - GM}{c^2 R})) & 0 & 0 \\ 0 & 0 & -R^2 & 0 \\ 0 & 0 & 0 & -R^2 sin^2\theta \end{pmatrix}$$

For an electrically neutral source, the metric becomes the Rosen metric. That metric is not a solution to the Einstein field equations, but has been studied in the literature because it is consistent with experimental tests of gravity to first order parameterized post-Newtonian (PPN) accuracy. At second order PPN accuracy, equation exponential_metric, predicts 0.7\microarcseconds more bending of star light by the Sun. An experiment such as LATOR, could confirm or reject this proposal on purely experimental grounds.

The bimetric theory of Rosen is not consistent with strong field tests of gravity. The fixed background metric can store momentum, allowing for bipolar modes of emission of gravity waves which are ruled out by observations of binary pulsars. The GEM action does not have an additional field to store momentum. For an isolated source, the lowest mode of emission that conserves momentum will be the quadrapole mode, consistent with observation.

QUANTIZATION

The 4D wave equation, has already been quantized. They were quantizing the field only to do the work of electromagnetism in a manifestly covariant way. The transverse modes of emission do the work of electromagnetism. The scalar mode of a spin 1 field caused a serious technical problem: it suggested negative probability states. To solve this issue, a supplementary condition was imposed to eliminate both the scalar and longitudinal modes.

There are two independent ways to spot the spin 1 field in the classical electrodynamic Lagrangian. The first is by looking at the anti-symmetric, rank two field strength tensor, $\nabla_{\mu} A_{\mu} - \nabla_{\mu} A_{\mu}$. If the indexes are swapped, then all the signs flip, the sign of an odd spin tensor.

The second approach considers current-current interactions of the charge coupling term, - $J^{\mu} A_{\mu}$. One can take the Fourier transformation of the 4-potential and write it as a current density in the momentum representation, - $(1/k^2) J' J^{\mu}$. If one constrains the current to move along the z axis, conservation of electric charge can be used to eliminate the current density. The current-current interaction is thus:

$$-\frac{1}{k^2} J_q^{'\mu} J_\mu = -\frac{1}{w^2} j'_\mu j_\mu + \frac{1}{w^2 - k^2}\left(j'_x j_x + j'_y j_y\right)$$

The current along the z axis is real. The currents along x and y are for the two independent polarizations of virtual photons needed for relativistic corrections. What is the symmetry of these virtual photons? Form the product of these two virtual currents:

$$\left(0, j_x', j_y', 0\right)\left(0, j_x, j_y, 0\right) = \left(j_x' j_z + j_y' j_y, 0, 0, j_y' j_z - j_z' j_y\right)$$

The phase part of this product (the z component) will require a 2π rotation to return its original location, a defining characteristic of a spin 1 field.

The GEM action has the antisymmetric field strength tensor, and this kind of current-current interaction. On a quantum level, the work of photons can be done.

The GEM action also has a spin 2 field. There is a symmetric, rank 2 field strength tensor, $\nabla_\mu A_\mu - \nabla_\mu A_\mu$. If the indexes are swapped, then all the signs stay the same, the sign of an even spin tensor. Since the tensor has indexes, it cannot be a scalar field. A spin 0 field can be formed by taking the trace of the symmetric field strength tensor, a Lorentz invariant quantity. If the trace is zero, then the particle can travel at the speed of light c. If the trace is not zero, then the scalar field formed will break gauge symmetry. In other words, a scalar inertial mass field breaks gauge symmetry. The Higgs mechanism is not necessary.

The analysis of the current-current density proceeds along lines similar to the case for electromagnetism. There is a difference for calculating the product of two currents. The real current is the one along z, yet the current along z does not appear in Equation. Although the reader may realise the currents along x and y in Equation are virtual, it would be preferable to enforce that they are zero. Working with pairs of currents with xz and yz, it may be possible to do both. The conjugate used in Equation is not the only anti-involutive automorphism. There is also a conjugate that flip all the signs but the first term of the 3-vector referred here as the first conjugate. One can also construct the second conjugate to keep the sign of the y current positive while flipping all others.

These products have the character of spin 2 because the two parts add together, allowing a return after only a ð rotation. If the three product currents are added together, The result has the real current along the z axis, and phases for spin 1 and spin 2 polarization modes. The four modes account for the four degrees of freedom in the 4-current. The scalar and longitudinal modes of emission will do the work of gravity. General relativity predicts transverse modes of emission for gravity waves. Should a gravitational wave detection experiment succeed, the polarization of the gravity wave will serve as an experimental test of this proposal.

There is a strong belief that field equations for gravity must be nonlinear. To be more precise, this means that gravity fields gravitate. The same is not true for electromagnetism: the electromagnetic field cannot be a source of charge. There are thought experiments to support the notion that gravity must be nonlinear. An example from a review paper on general relativity had two boxes with six neutral particles in each. Imagine that the rest mass of one particle were to be converted entirely into the kinetic energy of the other five. Would the first box be able to tell any difference in energy density and spacetime curvature between the two? If the answer is no, gravity must be governed by a nonlinear field equation.

Repeat the thought experiment, but this time make all the particles positively charged. Now the thought experiment cannot be done because it would involve destroying an electric charge. There are solid theoretical and experimental tests demonstrating that electric charge cannot be destroyed. Because the thought experiment cannot be done, the conclusion is not supported.

A NEW IMPLEMENTATION

The experimental tests of the equivalence principle indicate that gravity must be a metric theory. This paper questions how to implement a metric theory. General relativity takes a direct approach: start with the Riemann curvature tensor that characterizes how a metric changes, and put its contraction the Ricci scalar into the action. This paper uses the symmetry found in the definition of a covariant derivative to allow spacetime geometry to change based on what mass is in that spacetime. It is the flexibility of the spacetime geometry that indicates the approach can be characterized as a metric theory.

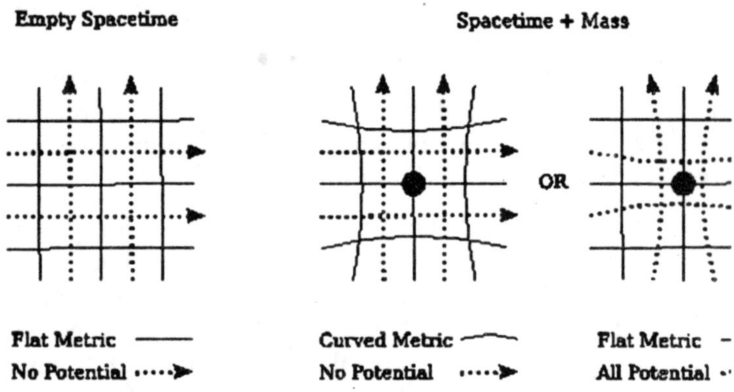

Fig. Metric or Potential for Gravity

There is a well-known thought experiment of a man in a closed box who would be unable to tell if he was in a smoothly accelerating rocket ship or the box was sitting on the surface of a planet ignoring tidal effects. For the box on the planet, general relativity would be able to generate a metric equation that in a limit process, would become a potential theory. For the GEM approach, one could choose a dynamic metric which would be equivalent to first order PPN accuracy to the metric from general relativity, or with the practical power of a diffeomorphism, choose to work in a flat spacetime and have a 4-potential completely characterize the gravitational field. A scalar potential like Newton's law of gravity is inadequate because there is only one parameter, so it gets the answer to light bending around the Sun half right. With four parameters, a 4-potential can describe the way measurements of both time and space are bent by a 4-momentum current density.

The GEM action is to a surprising degree exactly what Einstein searched for over the last half of his life. He tried an impressive range of ideas, from five dimensional Kaluza-Klein variations, to asymmetric Riemann curvature tensors, all to no avail. The radical approach used here is to abandon the Riemann curvature tensor, and work only with the building blocks of the curvature tensor, the connection as found in a covariant derivative. The 4D wave equation in this paper was called beautiful by Feynman in his lecture series. Elegance governs the heavens.

CLASSICAL UNIFIED FIELD EQUATIONS FOR GRAVITY AND ELECTROMAGNETISM

The equations governing gravity and electromagnetism show both profound similarities and unambiguous differences. Albert Einstein worked to unify gravity and electromagnetism, mainly by trying to generalize Riemannian geometry. Hamilton's quaternions are a 4-dimensional topological algebraic field related to the real and complex numbers equipped with a static Euclidean 4-basis. Riemannian quaternions as defined herein explicitly allow for dynamic changes in the basis vectors. The equivalence principle of general relativity which applies only to mass is generalized because for any Riemannian quaternion differential equation, the chain rule means a change could be caused by the potential and/or the basis vectors. The Maxwell equations are generated using a quaternion potential and operators. Unfortunately, the algebra is complicated. The unified force field proposed is modeled on a simplification of the electromagnetic field strength tensor, being formed by a quaternion differential operator acting on a potential, Box* A*. This generates an even, antisymmetric-matrix force field for electricity and an odd, antisymmetric-matrix force field for magnetism, where the even field conserves its sign if the order of the differential and the potential are reversed unlike the odd field. Gauge symmetry is broken for massive particles by the even, symmetric-matrix term, which is interpreted as being due to gravity. In tensor analysis, a differential operator acting on the field strength tensor creates the Maxwell equations. The unified field equations for an isolated source are generated by acting on the unified force field with an additional differential operator, Box* Box* A* = 4 pi J. This contains a quaternion representation of the Maxwell equations, a classical link to the quantum Aharonov-Bohm effect, and dynamic field equations for gravity. Vacuum and zero net current solutions to the unified field equations are discussed. The field equations conserve both electric charge density and mass density. Under a Lorentz transformation, the gravitational and electromagnetic fields are Lorentz invariant and Lorentz covariant respectively, but there are residual terms whose meaning is not clear presently. An additional constraint is required for gauge transformations of a massive field.

Einstein's Vision using Quaternions

Three of the four known forces in physics have been unified via the standard model:the electromagnetic, the weak, and the strong forces. The holdout remains gravity, the first force characterized mathematically by Isaac Newton. The parallels between gravity and electromagnetism are evident. Newton's law of gravity and Coulomb's law are inverse square laws. Both forces can be attractive, but Coulomb's law can also be a repulsive force. A longstanding goal of modern physics is to explain the similarities and differences between gravity and electromagnetism.

Albert Einstein had a specific idea for how to formulate an acceptable unified field theory taken from A. Pais, "Subtle is the Lord..." the science and life of Albert Einstein". One unusual aspect of Einstein's view was that he believed the unified field would lead to a new foundation for quantum mechanics, an idea which is not shared by some of today's thinkers. Most of Einstein's efforts over 40 years were directed in a search to generalize Riemannian differential geometry in four dimensions.

To a degree which has pleasantly surprised the author, Einstein's vision to unify gravity and electromagnetism has been followed. The construction of a new 4-dimensional geometry is dictated by insights garnered from physics. Events in spacetime are composed of a scalar for time and a 3-vector for space. The four-dimensional topological algebraic field of quaternions has the same structure, so quaternions will be the starting point of this effort.

Laws of physics are expressed in a coordinate-independent way. The sum or difference of two quaternions can only be defined if the two quaternions in question share the same 4-basis. Riemannian quaternions make coordinate-independence explicit. In special relativity, regions in spacetime are delimited by the light cone, where the net change in 3-space is equal to the net change in time. The parity between changes in 3-space and time is constructed into the definition of a Riemannian quaternion. In general relativity, the field equations make the metric a dynamic variable. The basis vectors of Riemannian quaternions can be dynamic, so the metric can be dynamic. The dynamic nature of the basis vectors leads to the general equivalence principle, whereby any law, even those in electromagnetism, can be the result of a change in reference frame.

Physical laws are the result of simple Riemannian quaternion differential equations. First-order Riemannian quaternion differential equations create force fields for gravity, electricity, and magnetism. Second-order differential equations create dynamic field equations for gravity, the Maxwell equations for electromagnetism, and a classical counterpart to the Aharanov-Bohm effect of quantum mechanics. Third-order differential equations create conservation laws. Homogeneous solutions to the second order differential equations are related to gauge symmetry.

The second paper in this series of three investigates a unified force law, with a focus on a particular solution which may eliminate the need for dark matter to explain the mass distribution and velocity profile for spiral galaxies. The third paper develops a new approach to quaternion analysis. The equations of the first two papers are recast with the new definition of a quaternion derivative, resulting in a quantum unified field and force theory.

Events in Spacetime and Quaternions

An event in spacetime is considered by the author as the fundamental form of information in physics. Events have structure. There are four degrees of freedom divided into two dissimilar parts:time is a scalar, and space is a 3-vector. This structure should be reflected in all the mathematics used to describe patterns of events. For this reason, this paper focuses exclusively on quaternions, the 4-dimensional number where the terms scalar and vector where first used.

Hamilton's quaternions, along with the far better know real and complex numbers, can be added, subtracted, multiplied and divided. Technically, these three numbers are the only finite-dimensional, associative, topological, algebraic fields, up to an isomorphism. Properties of these numbers are summarized in the table by dimension, if totally ordered, and if multiplication commutes:

Number	Dimensions	Totally Ordered	Commutative
Real	1	Yes	Yes
Complex	2	No	Yese
Quaternions	4	No	No

Hamilton's quaternions have a Euclidean 4-basis composed of 1, i, j, and k. The rules of multiplication were inspired by those for complex numbers: 1^2=1, i^2=j^2=k^2=ijk= –1. Quaternions also have a real 4 × 4 matrix representation:

$$q(t, x, y, z) = \begin{pmatrix} t & -x & -y & -z \\ x & t & -z & y \\ y & z & t & -x \\ z & -y & x & t \end{pmatrix}$$

Although written in Cartesian coordinates, quaternions can be written in any linearly-independent 4-basis because matrix algebra provides the necessary techniques for changing the basis. Therefore, like tensors, a quaternion equation is independent of the chosen basis. One could view quaternions as tensors restricted to a 4-dimensional algebraic field. For the sake of consistency, all transformations are also constrained to the same division algebra. This constraint might first appear too restrictive since for example it eliminates simple matrices for row permutations. Since quaternions are an algebraic field, there necessarily exists a combination of quaternions that achieves the action of a permutation. The need for consistency will overrule convenience.

Laws in physics are independent of coordinate systems. To make the coordinate independence explicit, amplitudes and basis vectors will be separated using a new notation. Consider a quaternion 4-function, A_n=(a_0, a_1, a_2, a_3), and an arbitrary 4-basis, Ihat_n = (ihat 0, ihat 1, ihat 2, ihat 3). In spacetime, the line that divides causality is define by the light cone. On the light cone, the total change in 3-space over the change in time is equal to one. Physics therefore indicates parity between the total 3-vector and the scalar, instead of weighing all four equally. A coordinate-independent Riemannian quaternion is defined to be A_0 Ihat_n=(a0 ihat0/3, a1 ihat1/3, a2 ihat2/3, a3 ihat3/3). The scaling factor of a third for the 3-vector plays a vital role in the definition of a regular function in the third paper of this series.

The equivalence principle of general relativity asserts, with experiments to back it up, that the inertial mass equals the gravitational mass. An accelerated reference frame can be indistinguishable from the effect of a mass density. No corresponding principle applies to electromagnetism, which depends only on the electromagnetic field tensor built from the potential. With Riemannian quaternions, the 4-unit vector does not have to be static, as illustrated by taking the time derivative of the first term and using the chain rule:

$$\frac{\partial a_0 \hat{i}_0}{\partial i_0} = \hat{i}_0 \frac{\partial a_0}{\partial i_0} + a_0 \frac{\partial \hat{i}_0}{\partial i_0}$$

The unit vector for time, ihat0, can change over an infinitely small amount of time, i0. Any change in a quaternion potential function could be due to contributions from a change in potential, the ihat0 da0/di0 term, and/or a change in the basis, the a0 dihat0/di0 term. Is this mathematical property related to physics? Consider Gauss' law written with Riemannian quaternions:

$$-\frac{\hat{i}_n^{\,2}}{9}\frac{\partial E_n}{\partial i_n} - \frac{\hat{i}_n E_n}{9}\frac{\partial \hat{i}_n}{\partial i_n} = 4\pi\rho, \quad n = 1, 2, 3$$

The divergence of the electric field might equal the source, or equivalently, the divergence of the basis vectors. The "general equivalence principle" as defined here means that any measurement can be due to a change in the potential and/or a change in the basis vectors. The general equivalence principle is applicable to both gravity and electromagnetism.

Metrics and Quaternion Products

The theories of special and general relativity dictate the distance between events in spacetime. Although fundamentally different in their mathematical structure, inertia is a link between the two. Special relativity dictates the transformation rules for observers who change their inertia, assuming the system observed does not change. The field equations of general relativity detail the changes in distance due to a system changing its inertia from the vacuum to a non-zero energy density. A quaternion product necessarily contains information about the metric, but also has information in the 3-vector. This additional information about quaternion products will suggest a provocative link between metrics and inertia consistent with both special and general relativity.

Most structures in Nature do not transform like a scalar and a 3-vector. Quaternion products multiply two 4-basis vectors, and those products will transform differently. The rules of quaternion multiplication mirror those of complex numbers. Instead of the imaginary number i, there is a unit 3-vector for each quaternion playing an analogous role. The difference is that unit 3-vectors do not all have to point in the same direction. Based on the angle between them, two different unit 3-vectors have both a dot and cross product. The dot and cross products completely characterize the relationship between the two unit vectors. Compare the product of multiplying two complex numbers (a, bi) and (c, di):

(a, bi) (c, di) = (ac − bd, ad + bc), with two quaternions, $\left(a, \vec{B}\right)$ and $\left(c, \vec{D}\right)$,

Complex numbers commute because they do not have a cross product in the result. If the order of quaternion multiplication is reversed, then only the cross product would change its sign. Quaternion multiplication does not commute due to the behaviour of the cross product. If the cross product is zero, then quaternion multiplication has all of the properties of complex numbers. If, on the other hand, the only value of a quaternion product is equal to the cross product, then multiplication is anti-commutative. Individually, the mathematical properties of commuting and anti-commuting algebras are well known. A quaternion product is the superposition of these two types of algebras that forms a division algebra.

Several steps are required to square of the difference of two Riemannian quaternions to form a measure of distance. First, the basis of the two quaternions must be shared. It makes no sense to subtract something in spherical coordinates from something in Cartesian coordinates. The basis does not have to be constant, only shared. Every quaternion commutes with itself, so the cross product is zero. There are seven unique pairs of basis vectors in a square:

$$\left(da_0\hat{i}_0, dA_n\frac{\hat{I}_n}{3}\right)^2 = \left(da_0^2\hat{i}_0^2\frac{\hat{I}_n^2}{9}, 2da_0dA_n\frac{\hat{I}_0\hat{I}_n}{3}\right)$$

The signs were chosen to be consistent with Hamilton's quaternion algebra. The four square basis vectors i_mu^2 define the metric. If the basis vectors are not constant, then the metric is dynamic. Define a "3-rope" to be the three other terms, which have the form i0 In. Notice that the 3-rope starts in one time-space location and will have a non-zero length if it ends up at a different location and time. With quaternion products, the 3-rope is a natural companion to a metric for information about distance.

In special relativity, if the inertia of the observer but not the system is changed, the metric is invariant. The 3-rope is covariant, because it is known how it changes. Given the utility of duality, a complementary hypothesis to the invariant metric of special relativity would propose the if the inertia of the system but not the observer is changed, there exists a choice of basis vectors such that the 3-rope is invariant but the metric changes in a known way. This could be written in algebraically using the following rule:

$$\hat{i}_0^{\,2} = \frac{-1}{\hat{I}_n^{\,2}}, \quad | \, \hat{i}_0 \, \hat{I}_n \, | = 1$$

If the magnitude of the time and 3-space basis vectors are inversely related, the magnitude of the product of the time basis vector with each 3-space basis vector will be constant even if the basis vectors themselves are dynamic. This hypothesis asserts the existence of such a basis, but that particular basis does not have to be used.

Hamilton had the freedom to use the rule found in the above equation, but made the more obvious choice of i_0^2 = -I_n^2. The existence of a basis where the 3-rope is constant despite a change in the inertia of the system will have to be treated as provisional in this paper. In the second paper of this series, a metric with this property will be found and discussed.

Physically Relevant Differential Equations

Is there a rational way to construct physically relevant quaternion equations? The method used here will be to mimic the tensor equations of electromagnetism. The electromagnetic field strength tensor is formed by a differential operator acting on a potential. The Maxwell equations are formed by acting on the field with another differential operator. The Lorentz 4-force is created by the product of a electric charge, the electromagnetic field strength tensor, and a 4-velocity. This pattern will be repeated starting from an asymmetric field to create the same field and force equations using quaternion differentials and potentials. The challenge in this exercise is in the interpretation, to see how every term connects to established laws of physics.

As a first step to constructing differential equations, examine how the differential operator (d/dt, Del) acts on a potential function (phi, A):

$$\left(\frac{\partial}{\partial t}, \vec{\nabla}\right)(\phi, \vec{A}) = \left(\frac{\partial \phi}{\partial t} - \vec{\nabla}.\vec{A},\ \frac{\partial \vec{A}}{\partial t} + \vec{\nabla}\phi + \vec{\nabla} \times \vec{A}\right)$$

For the sake of clarity, the notation introduced for Riemann quaternions has been suppress, so the reader is encouraged to recognize that there are also a parallel set of terms for changes in the basis vectors. The previous equation is a complete assessment of the change in the 4-dimensional potential/basis, involving two time derivatives, the divergence, the gradient and the curl all in one. A unified field theory should account for all conceivable forms of change in a 4-dimensional potential/basis, as is the case here.

Quaternion operators and potentials have not been used to express the Maxwell equations. The reason can be found in the previous equation, where the sign of the divergence of A is opposite of the curl of A. In the Maxwell equations, the divergence and the curl involving the electric and magnetic field are all positive. Many others, even in Maxwell's time, have used complex-valued quaternions for the task because the extra imaginary number can be used to get the signs correct. However, complex-valued quaternions are not an algebraic field. The norm, t^2+x^2+y^2+z^2, for a non-zero quaternion could equal zero if the values of t, x, y, and z were complex. This paper involves the constraint of working exclusively with 4-dimensional algebraic fields. Therefore, no matter how salutary the work with complex-valued quaternions, it is not relevant to this paper.

The reason to hope for unification using quaternions can be found in an analysis of symmetry provided by Albert Einstein:

"The physical world is represented as a four-dimensional continuum. If in this we adopt a Riemannian metric, and look for the simplest laws which such a metric can satisfy, I arrive at the relativistic gravitation theory of empty space. If we adopt in this space a vector field, or the anti symmetric tensor field derived from it, and if we look for the simplest laws which such a field can satisfy, I arrive at the Maxwell equations for free space."

The "four-dimensional continuum" could be viewed as a technical constraint involving topology. Fortunately, quaternions do have a topological structure since they have a norm. Nature is asymmetric, containing both a symmetric metric for gravity and an antisymmetric tensor for electromagnetism. With this in mind, rewrite out the real 4x4 matrix representation of a quaternion:

$$q(t, x, y, z) = \begin{pmatrix} t & 0 & 0 & 0 \\ 0 & t & 0 & 0 \\ 0 & 0 & t & 0 \\ 0 & 0 & 0 & t \end{pmatrix} + \begin{pmatrix} 0 & -x & -y & -z \\ x & 0 & -z & y \\ y & z & 0 & -x \\ z & -y & x & 0 \end{pmatrix}$$

The scalar component can be represented by a symmetric 4 × 4 matrix, invariant under transposition and conjugation (these are the same operations for quaternions). The 3-vector component (x, y and z in the representation above) is off-diagonal and can be represented by an anti symmetric 4x4 matrix, because taking the transpose will flip the signs of the 3-vector. Quaternions are asymmetric in their matrix representation, a property which is critical to using them for unifying gravity and electromagnetism.

Recreating the Maxwell Equations

Maxwell speculated that his set of equations might be expressed with quaternions someday. The divergence, gradient, and curl were initially developed by Hamilton during his investigation of quaternions. For the sake of logical consistency, any system of differential equations, such as the Maxwell equations, that depends on these tools must have a quaternion representation.

The Maxwell equations are gauge invariant. How can this property be built into a quaternion expression? Consider a common gauge such as the Lorenz gauge, dphi/dt + div A = 0. In quaternion parlance, this is a quaternion-scalar formed from a differential quaternion acting on a potential. To be invariant under an arbitrary gauge transformation, the quaternion-scalar must be set to zero. This can be done with the vector operator, (q – q*)/2. Search for a combination of quaternion operators and potentials that generate the Maxwell equations:

$$\frac{-(\square \text{Vector}(\square^* A^*) + \square^* \text{Vector}(\square A))^*}{2} =$$
$$= \left(\vec{\nabla} \cdot \left(-\frac{\partial \vec{A}}{\partial t} - \vec{\nabla}\phi\right), \frac{\partial \vec{\nabla} \times \vec{A}}{\partial t} + \vec{\nabla} \times \left(-\frac{\partial \vec{A}}{\partial t} - \vec{\nabla}\phi\right)\right) =$$
$$= \left(\vec{\nabla} \cdot \vec{E}, \frac{\partial \vec{B}}{\partial t} + \vec{\nabla} \times \vec{E}\right) =$$
$$= \left(4\pi\rho, \vec{0}\right).$$

This is Ampere's law and the no monopoles vector identity assuming a simply-connected topology. Any choice of gauge will not make a contribution due to the vector operator. If the vector operator was not used, then the gradient of the symmetric-matrix force field would be linked to the electromagnetic source equation, Ampere's law.

Generate the other two Maxwell equations:

This is Gauss' and Faraday's law. Again, if the vector operator had not been used, the time derivative of the symmetric-matrix force field would be associated with the electromagnetic source equation, Gauss' law. To specify the Maxwell equations completely, two quaternion equations are required, just like the 4-vector approach.

Although successful, the quaternion expression is unappealing for reasons of simplicity, consistency and completeness. A complicated collection of sums or differences of differential operators acting on potentials - along with their conjugates - is required. There is no obvious reason this combination of terms should be central to the nature of light. One motivation for the search for a unified potential field involves simplifying the above expressions.

When a quaternion differential acts on a function, the divergence always has a sign opposite the curl. The opposite situation applies to the Maxwell equations. Of course the signs of the Maxwell equations cannot be changed. However, it may be worth the effort to explore equations with sign conventions consistent with the quaternion algebra, where the operators for divergence and curl were conceived.

Information about the change in the potential is explicitly discarded by the vector operator. Justification comes from the plea for gauge symmetry, essential for the Maxwell equations. The Maxwell equations apply to massless particles. Gauge symmetry is broken for massive fields. More information about the potential might be used in unification of electromagnetism with gravity. A gauge is also matrix symmetric, so it could provide a complete picture concerning symmetry.

One unified force Field from one Potential Field

For massless particles, the Maxwell equations are sufficient to explain classical and quantum electrodynamic phenomena in a gauge-invariant way. To unify electromagnetism with gravity, the gauge symmetry must be broken, opening the door to massive particles. Because of the constraints imposed by quaternion algebra, there is little freedom to choose the gauge with a simple quaternion expression. In the standard approach to the electromagnetic field, a differential 4-vector acts on a 4-vector potential in such a way as to create an antisymmetric second-rank tensor. The unified field hypothesis proposed involves a quaternion differential operator acting on a quaternion potential:

$$\Box^* A^* = \left(\frac{\partial \phi}{\partial t} - \vec{\nabla} . \vec{A}, \; - \frac{\partial \vec{A}}{\partial t} - \vec{\nabla} \phi + \vec{\nabla} \times \vec{A} \right)$$

This is a natural suggestion with this algebra. The antisymmetric-matrix component of the unified field has the same elements as the standard electromagnetic field tensor. Define the electric field E as the even terms, the ones that will not change signs if the order of the differential operator and the potential are reversed. The magnetic field B is the curl of A, the odd term. The justification for proposing the unified force field hypothesis rests on the presence of the electric and magnetic fields.

In some ways, the above equation looks just like the old idea of combining a scalar gauge field with the electromagnetic field strength tensor, as Gupta did in 1950 in order to quantize the Maxwell equations. He concluded that although useful because it is written in manifestly relativistic form, no new results beyond the Maxwell equation are obtained. Examine just the gauge contribution to the Lagrangian for this unified field:

$$L = -\frac{1}{2}\left(\frac{\partial \phi}{\partial t}, -\frac{1}{3}\frac{\partial A_x}{\partial x}, -\frac{1}{3}\frac{\partial A_y}{\partial y}, -\frac{1}{3}\frac{\partial A_z}{\partial z}\right)^2$$

Take the derivative of the Lagrangian with respect to the gauge variables:

$$\frac{\partial L}{\partial \frac{\partial A_\mu}{\partial x_\mu}} = 0$$

By Noether's theorem, this conserved current indicates a symmetry of the Lagrangian. This is why the proposal involves new physics. The gauge is a dynamic variable constrained by the Lagrangian.

A quaternion potential function has four degrees of freedom represented by the scalar function phi and the 3-vector function A. Acting on this with one differential operators does not change the degrees of freedom. Instead, the tangent spaces of the potential will offer more subtle views on the rules for how potentials change.

The three classical force fields, g, E, and B, depend on the same quaternion potential, so there are only four degrees of freedom. With seven components to the three classical force fields, there must be three constraints between the fields. Two constraints are already familiar. The electric and magnetic field form a vector identity via Faraday's law. Assuming spacetime is simply connected, the no monopoles equation is another identity. A new constraint arises because both the force fields for gravity and electricity are even. It will be shown subsequently how the even force fields can partially constructively or destructively interfere with each other.

Unified Field Equations

In the standard approach to generating the Maxwell equations, a differential operator acts on the electromagnetic field strength tensor. A unified field hypothesis for an isolated source is proposed which involves a differential quaternion operator acting on the unified field:

$$4\pi(\rho, \vec{J})^* = \left(\frac{\partial}{\partial t}, \vec{\nabla}\right)^* \left(\frac{\partial}{\partial t}, \vec{\nabla}\right)^* (\phi, \vec{A})^* =$$

$$= \left(\frac{\partial^2 \phi}{\partial t^2} - 2\vec{\nabla}\cdot\frac{\partial \vec{A}}{\partial t} - \vec{\nabla}\cdot\vec{\nabla}\phi + \vec{\nabla}\cdot(\vec{\nabla}\times\vec{A}),\right.$$

$$-2\vec{\nabla}\frac{\partial \phi}{\partial t} + \vec{\nabla}(\vec{\nabla}\cdot\vec{A}) - \frac{\partial^2 \vec{A}}{\partial t^2} - \vec{\nabla}\times(\vec{\nabla}\times\vec{A}) +$$

$$\left. +2\vec{\nabla}\times\frac{\partial \vec{A}}{\partial t} + \vec{\nabla}\times\vec{\nabla}\phi\right).$$

This second order set of four partial differential equations has four unknowns so this is a complete set of field equations. Rewrite the equations above in terms of the classical force fields:

$$4\pi(\rho, \vec{J})^* = \left(\frac{\partial g}{\partial t} + \vec{\nabla}.\vec{E} + \vec{\nabla}.\vec{B},\right.$$

$$\left. -\vec{\nabla}g + \frac{\partial \vec{E}}{\partial t} - \vec{\nabla}\times\vec{B} + \frac{\partial \vec{B}}{\partial t} - \vec{\nabla}\times\vec{E}\right).$$

The unified field equations contain three of the four Maxwell equations explicitly:Gauss' law, the no magnetic monopoles law, and Ampere's law. Faraday's law is a vector identity, so it is still true implicitly. Therefore, a subset of the unified field equations contains a quaternion representation of the Maxwell equations. The justification for investigating the unified field equation hypothesis is due to the presence of the Maxwell equations.

There is a very simple relationship between Faraday's law and the equation above. All that needs to be done is to subtract twice the time derivative of the magnetic field from both sides. What does this do to the 4-vector current density J? Now there is a current that transforms like a pseudo-current density, making the proposal more complete. The volume integral of this pseudo-current density is the total magnetic flux:

The unified field equation postulates a pseudo 3-vector current composed of the difference between the time derivative of the magnetic field and the curl of the electric field. The Aharonov-Bohm effect depends on the total magnetic flux to create changes seen in the energy spectrum.The volume integral of the time derivative of the magnetic field is a measure of the total magnetic flux. The pseudo-current density is quite unusual, transforming differently under space inversion than the electric current density. One might imagine that a Lorentz transformation would shift this pseudo-current density into a pseudo-charge density. This does not happen however, because the vector identity involving the divergence of a curl still applies. The Aharonov-Bohm phenomenon, first viewed as a purely quantum effect, may have a classical analogue in the unified field equations.

$$k \iiint \frac{\partial B}{\partial t}\, dV = \frac{e}{\hbar c}\, \Phi_B$$

The constants used here were chosen to suggest a connection to the Aharonov-Bohm. The Aharonov-Bohm effect depends on the total magnetic flux to create changes seen in the energy spectrum. A pseudo-current density is unusual, transforming differently under space inversion than the electric current density. One might imagine that a Lorentz transformation would shift this pseudo-current density into a pseudo-charge density. This does not happen however, because the vector identity involving the divergence of a curl still applies. The Aharonov-Bohm phenomenon, first viewed as a purely quantum effect, may have a classical analogue in the unified field equations.

The field equations involving the gravitational force field are dynamic and depend on four dimensions. This makes them likely to be consistent with special relativity. Since they are generated alongside the Maxwell equations, one can reasonably expect the differential equations will share many properties, with the ones involving the symmetric-matrix gravitational force field being more symmetric than those of the electromagnetic counterpart.

The unified source can be defined in terms of more familiar charge and current densities by separately setting the gravity or electromagnetic field equal to zero. In these cases, the source is due only to electricity or mass respectively. This leads to connections between the unified source, mass, and charge:

$$J = J_m \text{ iff } \vec{E} = \vec{B} = \vec{0}$$
$$J = J_e + \vec{J}_{AB} \text{ iff } g = 0.$$

It would be incorrect - but almost true - to say that the unified charge and current are simply the sum of the three: mass, electric charge, and the Aharanov-Bohm pseudo-current. These terms constructively interfere with each other, so they may not be viewed as being linearly independent.

Up to four linearly independent unified field equations can be formulated. A different set could be created by using the differential operator without taking its conjugate:

$$-4\pi J = \Box^* \Box^* A^* =$$
$$= \left(\frac{\partial^2 \phi}{\partial t^2} + \vec{\nabla} \cdot \vec{\nabla}\phi - \vec{\nabla} \cdot (\vec{\nabla} \times \vec{A}), -\frac{\partial^2 \vec{A}}{\partial t^2} - \vec{\nabla}^2 \vec{A} - \vec{\nabla} \times \right.$$
$$= \left(\frac{\partial g}{\partial t} - \vec{\nabla}.\vec{E} - \vec{\nabla}.\vec{B}, \right.$$
$$\left. \vec{\nabla} g + \frac{\partial \vec{E}}{\partial t} + \vec{\nabla} \times \vec{B} + \frac{\partial \vec{B}}{\partial t} + \vec{\nabla} \times \vec{E} \right).$$

This is an elliptic equation. Since the goal of this work is a complete system of field equations, this may turn out to be an advantage. An elliptic equation combined with a hyperbolic one might more fully describe gravitational and electromagnetic waves from sources. Unlike the first set of field equations, the cross terms destructively interfere with each other.

The elliptic field equation again contains three of four Maxwell equations explicitly:Gauss' law, the no magnetic monopoles vector identity and Faraday's law. This time, Ampere's law looks different. To be consistent with Ampere's law, again a pseudo-current must be included. This may be the differential form of a classical Aharonov-Bohm effect.

The only term that does not change between the two field equations is the one involving the dynamic gravitational force. This might be a clue for why this force is only attractive.

Solutions to the Unified Field Equations

All the solutions that have been worked out for the Maxwell equations will work with the unified field equations. For example, if the potential is static, the scalar equation for hyperbolic field equation is the Poisson equation. The unified equations are more informative, since any potential which is a solution to the scalar Poisson equation will also characterize the corresponding current.

The field equations of general relativity and the Maxwell equations both have vacuum solutions,such as plane wave solutions. The unified field equations do not have such a solution, other than a constant. Given historical tradition, this may seem like a deadly flaw. However, it may be something that is required for a final and complete theory. In a unified field theory, the gravitational part may be zero while the electrical part is not, and visa versa. Non-zero solutions are worth exploring

An inverse square potential plays an important role in both gravity and electromagnetism. Examine the scalar field involving the inverse interval squared:

$$\Box\Box^*\left(\frac{1}{t^2 - x^2 - y^2 - z^2}, \vec{0}\right) = \left(\frac{4\,(3\,t^2 + x^2 + y^2 + z^2)}{(t^2 - x^2 - y^2 - z^2)^3}, \vec{0}\right)$$

This potential solves the Maxwell equations in the Lorentz gauge:

$$\Box^2 \left(\frac{1}{t^2 - x^2 - y^2 - z^2}, \vec{0} \right) = 0$$

The non-zero part may have everything to do with gravity.

A plane wave solution does exist, but not for a pure vacuum. Instead, a plane wave solution exists with the constrain that the net current is zero for the elliptical field equations.

The field equations of general relativity and the Maxwell equations both have vacuum solutions. A vacuum solution for the unified field equation is apparent for the elliptical field equations:

$$A = \left(\phi_0 \, e^{\vec{K} \cdot \vec{R} - \omega t}, \vec{A}_0 \, e^{\vec{K} \cdot \vec{R} - \omega t} \right)$$

The unified field equation will evaluate to zero if

$$\text{Scalar} \left(\left(\frac{\omega}{c}, \vec{K} \right)^2 \right) = 0$$

The dispersion relation is an inverted distance, so it will depend on the metric. The same potential can also solve the hyperbolic field equations under different constraints and resulting dispersion equation. There were two reasons for not including the customary imaginary number "i" in the exponential of the potential. First, it was not necessary. Second, it would have created a complex-valued quaternion, and therefore is outside the domain of this paper. The important thing to realise is that vacuum solutions to the unified field equations exist whose dispersion equations depend on the metric. This is an indication that unifying gravity and electromagnetism is an appropriate goal.

CONSERVATION LAWS

Conservation of electric charge is implicit in the Maxwell equations. Is there also a conserved quantity for the gravitational field?

Notice that the gravitational force field only appears in the quaternion scalar. The electromagnetic fields only appear in the 3-vector. This generates two types of constraints on the sources. No change in the electric source applies to the quaternion scalar. No change in the gravitational source applies to the 3-vector.

$$\text{Scalar } (\square J_e^*) = \frac{\partial \rho_e}{\partial t} + \vec{\nabla} \cdot \vec{J}_e = 0$$

$$\text{Scalar } \left(\square \vec{J}_{AB}^*\right) = \nabla \cdot \vec{J}_{AB}^* = 0$$

$$\text{Vector } (\square J_m^*) = -\frac{\partial \vec{J}_m}{\partial t} + \vec{\nabla} \rho_m - \vec{\nabla} \times \vec{J}_m = \vec{0}$$

The first equation is known as the continuity equation, and is the reason that electric charge is conserved. For a different inertial observer, this will appear as a conservation of electric current density. There is no source term for the Aharanov-Bohm current, and subsequently no conservation law. The 3-vector equation is a constraint on the mass current density, and is the reason mass current density is conserved. For a different inertial observer, the mass density is conserved.

TRANSFORMATIONS OF THE UNIFIED FORCE FIELD

The transformation properties of the unified field promise to be more intricate than either gravity or electromagnetism separately. What might be expected to happen under a Lorentz transformation? Gravity involves mass that is Lorentz invariant, so the field that generates it should be Lorentz invariant. The electromagnetic field is Lorentz covariant. However, a transformation cannot do both perfectly. The reason is that a Lorentz transformation mixes a quaternion scalar with a 3-vector. If a transformation left the quaternion scalar invariant and the 3-vector covariant, the two would effectively not mix. The effect of unification must be subtle, since the transformation properties are well known experimentally.

Consider a boost along the x-axis. The gravitational force field is Lorentz invariant. All the terms required to make the electromagnetic field covariant under a Lorentz transformation are present, but covariance of the electromagnetic fields requires the following residual terms:

$$(\Box'^{*}A'^{*})_{Residual} = \left(0,\ (\gamma^2\beta^2 - 1)\frac{\partial A_x}{\partial t} + (\gamma^2 - 1)\frac{\partial \phi}{\partial x},\right.$$
$$\left.-2\gamma\beta\left(\frac{\partial A_z}{\partial t} + \frac{\partial A_y}{\partial x}\right),\ 2\gamma\beta\left(\frac{\partial A_y}{\partial t} - \frac{\partial A_z}{\partial x}\right)\right).$$

At this time, the correct interpretation of the residual term is unclear. Most importantly, it was shown earlier that charge is conserved. These terms could be a velocity-dependent phase factor. If so, it might provide a test for the theory.

The mechanics of the Lorentz transformation itself might require careful re-examination when so strictly confined to quaternion algebra. For a boost along the x-axis, if only the differential transformation is in the opposite direction, then the electromagnetic field is Lorentz covariant with the residual term residing with the gravitational field. The meaning of this observation is even less clear. Only relatively recently has DeLeo been able to represent the Lorentz group using real quaternions. The delay appears odd since the interval of special relativity is the scalar of the square of the difference between two events. In the real 4 × 4 matrix representation, the interval is a quarter of the trace of the square. Therefore, any matrix with a trace of one that does not distort the length of the scalar and 3-vector can multiply a quaternion without effecting the interval. One such class is 3-dimensional, spatial rotations. An operator that adds nothing to the trace but distorts the lengths of the scalar and 3-vector with the constraint that the difference in lengths is constant will also suffice. These are boosts in an inertial reference frame. Boosts plus rotations form the Lorentz group.

Three types of gauge transformations will be investigated: a scalar, a 3-vector, and a quaternion gauge field. Consider an arbitrary scalar field transformation of the potential:

$$A \to A' = A - \Box^{*}\lambda.$$

The electromagnetic fields are invariant under this transformation. An additional constraint on the gauge field is required to leave the gravitational force field invariant, namely that the scalar gauge field solves a homogeneous elliptical equation. From the perspective of this proposal, the freedom to choose a scalar gauge field for the Maxwell equations is due to the omission of the gravitational force field.

Transform the potential with an arbitrary 3-vector field:

$$A \rightarrow A' = A - \square^* \vec{\Lambda}.$$

This time the gravitational force field is invariant under a 3-vector gauge field transformation. Additional constraints can be placed on the 3-vector gauge field to preserve a chosen electromagnetic invariant. For example, if the difference between the two electromagnetic fields is to remain invariant, then the 3-vector gauge field must be the solution to an elliptical equation. Other classes of invariants could be examined.

The scalar and 3-vector gauge fields could be combined to form a quaternion gauge field. This gauge transformation would have the same constraints as those above to leave the fields invariant. Is there any such gauge field? The quaternion gauge field can be represented the following way:

$$A \rightarrow A' = A - \square^* \Lambda.$$

If a force field is created by hitting this gauge transformation with a differential operator, then the gauge field becomes a unified field equation. Since vacuum solutions have been found for those equations, a quaternion gauge transformation can leave the field invariant.

FUTURE DIRECTIONS

The fields of gravity and electromagnetism were unified in a way consistent with Einstein's vision, not his technique. The guiding principles were simple but unusual:generate expressions familiar from electromagnetism using quaternions, striving to interpret any extra terms as being due to gravity. The first hypothesis about the unified field involved only a quaternion differential operator acting on a potential, no extra terms added by hand. It contained the typical potential representation of the electromagnetic field, along with a symmetric-matrix force field for gravity. The second hypothesis concerned a unified field equation formed by acting on the unified field with one more differential operator. All the Maxwell equations are included explicitly or implicitly. Additional terms suggested the inclusion of a classical representation of the Aharanov-Bohm effect. Four linearly independent unified field equations exist, but only the hyperbolic and elliptic cases were discussed. A large family of vacuum solutions exists, and will require future analysis to appreciate. To work within the guidelines of this paper, one should avoid solutions represented by complex-valued quaternions.

Why did this approach work? The hypothesis that initiated this line of research was that all events in spacetime could be represented by quaternions, no matter how the events were generated. This is a broad hypothesis, attempting to reach all areas covered by physics. Based on the equations presented in this paper, a logical structure can be constructed, starting from events. A set of events forms a pattern that can be described by a potential. The change in a potential creates a field. The change in field creates a field equation. The terms that do not change under differentiation of a field equation form conservation laws.

Chapter 5

Magnetic Materials

Magnetic materials encompass a wide variety of materials, which are used in a diverse range of applications. Magnetic materials are utilised in the creation and distribution of electricity, and, in most cases, in the appliances that use that electricity. They are used for the storage of data on audio and video tape as well as on computer disks. In the world of medicine, they are used in body scanners as well as a range of applications where they are attached to or implanted into the body. The home entertainment market relies on magnetic materials in applications such as PCs, CD players, televisions, games consoles and loud speakers.

It is difficult to imagine a world without magnetic materials and they are becoming more important in the development of modern society. The need for efficient generation and use of electricity is dependent on improved magnetic materials and designs. Non-polluting electric vehicles will rely on efficient motors utilising advanced magnetic materials. The telecommunications industry is always striving for faster data transmission and miniaturisation of devices, both of which require development of improved magnetic materials.

Magnetic materials are classified in terms of their magnetic properties and their uses. If a material is easily magnetised and demagnetised then it is referred to as a soft magnetic material, whereas if it is difficult to demagnetise then it is referred to as a hard (or permanent) magnetic material.

Materials in between hard and soft are almost exclusively used as recording media and have no other general term to describe them. Other classifications for types of magnetic materials are subsets of soft or hard materials, such as magnetostrictive and magnetoresistive materials.

HISTORY OF MAGNETIC MATERIALS

The history of permanent magnetism dates back over many centuries. The earliest observations of magnetism can be traced back to the Greek philosopher Thales in the 6th Century B.C. However, it was not until 1600 that the modern understanding of magnetism began.

1600: Dr. William Gilbert published the first systematic experiments on magnetism in *"De Magnete"*.

1819: Oerstead accidentally made the connection between magnetism and electricity discovering that a current carrying wire deflected a compass needle.

1825: Sturgeon invented the electromagnet.

1880: Warburg produced the first hysteresis loop for iron.

1895: The Curie law was proposed.

1905: Langevin first explained the theory of diamagnetism and paramagnetism.

1906: Weiss proposed ferromagnetic theory.

1920's: The physics of magnetism was developed with theories involving electron spins and exchange interactions; the beginnings of quantum mechanics.

The progress of permanent magnetic materials proceeds in a series of steps. Each material is developed and improved before being supplanted by a new one. This trend is shown in figure which shows the development of permanent magnet materials throughout the 20th century. The graph shows the improvement in maximum energy product of the materials, which is used as a figure of merit, as it is a measure of the ability of magnets to do work per unit volume of material. The reduction in size of magnets is illustrated more vividly in figure.

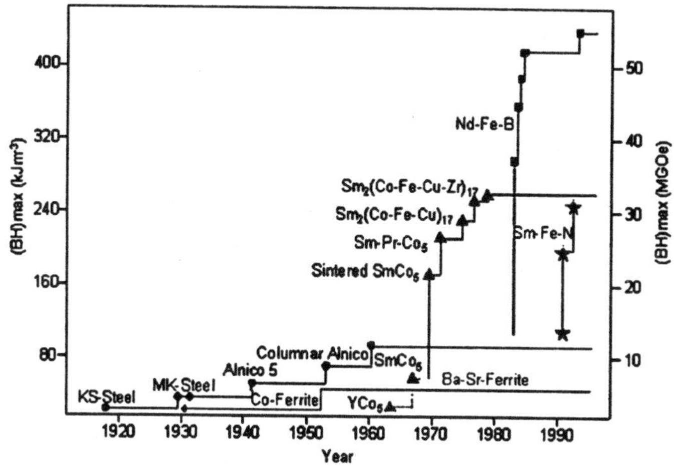

Fig. The development of permanent magnets in the 20th Century. BH_{max} has improved exponentially, doubling every 12 years.

Fig. Brass bound lodestone, ferrite block and NdFeB magnet: each store the same magnetic energy (~0.4J) & contain ~70percent iron by weight, yet the mass has decreased a thousand fold.

Lodestone

This was the first permanent magnet to be recognised; a naturally occurring oxide Fe_3O_4. Lightening strikes are thought to be responsible for the large magnetic fields that initially magnetised the rocks. The magnetic field produced is low, however, the resistance to demagnetisation is reasonably high.

Magnetic Carbon Steel

These were developed in the eighteenth century. These steels are normally alloyed with tungsten and/or chromium to form carbide precipitates under appropriate heat treatments, which are effective in obstructing domain wall movement. These magnets have a high magnetic saturation, far superior to lodestone, however, they are prone to demagnetisation necessitating the use of long shapes to minimise demagnetisation fields.

Alnico Magnets: (alloys based on Al, Co, and Ni)

This group of magnets developed in the 1930's were the first *modern* permanent magnets offering considerable in magnetic hardness over the magnetic steels previously available. Their properties rely on the shape anisotropy associated with the two phase nanostructure comprising of ferromagnetic Fe-Co needles in a matrix of non-magnetic Al-Ni. Due to their high Curie temperature, *~850°C*, they are still used for certain applications today.

Cobalt Platinum Magnets

These were developed in the 1950's. Their improved properties over the Alnico's and corrosion resistance made them an ideal candidate for use in biomedical applications at the time. However, due to their high cost they did not gain widespread use and the rare-earth magnets superseded them.

Hard Ferrite Magnets: ($BaFe_{12}O_{19}$ or $SrFe_{12}O_{19}$)

These have been of most important commercial magnets for the last few decades. Due to their anisotropic structure they exhibit relatively high coercivity, however, the energy product is low. Despite this they have found widespread use, as the abundance of the raw materials is high and production costs are low they are also suitable for use in complex shapes. As a result they still remain the most common magnetic material in bulk applications.

Samarium Cobalt

Developed in the late 1960's this group of alloys combine cobalt, iron and a light rare earth element; many exhibit high energy hard magnetic behaviour, but $SmCo_5$ is as yet the only one of commercial significance. These held the record for highest energy product for many years, unfortunately they have the disadvantage of being costly. These magnets have good thermal stability and are therefore used in applications where where the magnets will be exposed to high temperatures.

Neodymium Iron Boron Magnet

These were first produced in 1984; they combine a high saturation magnetisation with good resistance to demagnetisation. The high cost of samarium and the instability in the price of cobalt has lead to these magnets becoming the material of choice for applications requiring high energy magnets. Despite the high energy product these magnets have a relatively low Curie temperature (*312*°C) which limits their use in high temperature applications: additions of Co and Dy improves the temperature characteristics but also increases the production costs. This has not, however, prevented them being used increasingly in the marketplace, especially in applications where miniaturisation is an important design criteria.

Samarium Iron Nitrides

The development of these alloys is still on going; they are a promising new candidate for permanent magnet applications due to their high resistance to demagnetisation, high magnetisation and increased resistance to corrosion and temperature when compared with neodymium iron boron.

ORIGIN OF MAGNETISM

Almost everyone is familiar with what a magnetic material can do but very few know how a magnet works. To understand this phenomenon one must first grasp the inextricable connection that exists between magnetism and electricity.

A simple electromagnet can be produced by wrapping copper wire into the form of a coil and connecting the wire to a battery. A magnetic field is created in the coil but it remains there only while electricity flows through the wire.

An ordinary bar magnet does not have an obvious connection with electricity so how does it work? The field created by the magnet is associated with the motions and interactions of its electrons, the minute charged particles which orbit the nucleus of each atom.

Electricity is the movement of electrons, whether in a wire or in an atom, so each atom represents a tiny permanent magnet in its own right. The circulating electron produces its own orbital magnetic moment, measured in Bohr magnetons (μB), and there is also a spin magnetic moment associated with it due to the electron itself spinning, like the earth, on its own axis. In most materials there is resultant magnetic moments, due to the electrons being grouped in pairs causing the magnetic moment to be cancelled by its neighbour.

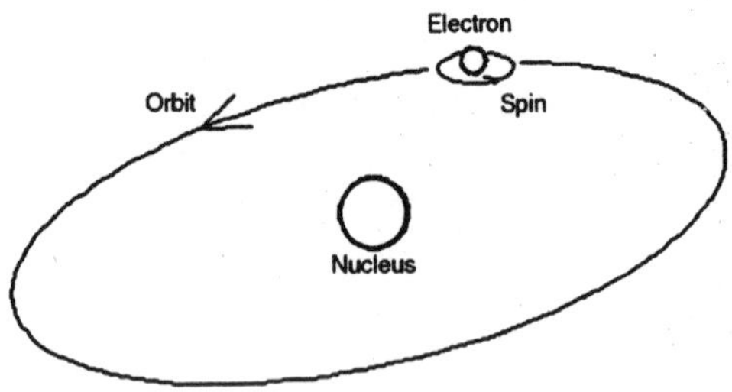

Fig. The orbit of a Spinning Electron about the Nucleus of an Atom.

In certain magnetic materials the magnetic moments of a large proportion of the electrons align, producing a unified magnetic field. The field produced in the material (or by an electromagnet) has a direction of flow and any magnet will experience a force trying to align it with an externally applied field, just like a compass needle. These forces are used to drive electric motors, produce sounds in a speaker system, control the voice coil in a CD player, etc...

The interactions between magnetism and electricity are therefore an essential aspect of many devices we use every day.

MAGNETIC UNITS & TERMINOLOGY

In the study of magnetism there are two systems of units currently in use: the mks (metres-kilograms-seconds) system, which has been adopted as the S.I. units and the cgs (centimetres-grams-seconds) system, which is also known as the Gaussian system. The cgs system is used by many magnets experts due to the numerical equivalence of the magnetic induction (*B*) and the applied field (*H*).

When a field is applied to a material it responds by producing a magnetic field, the magnetisation (*M*). This magnetisation is a measure of the magnetic moment per unit volume of material, but can also be expressed per unit mass, the specific magnetisation (*s*). The field that is applied to the material is called the applied field (*H*) and is the total field that would be present if the field were applied to a vacuum. Another important parameter is the magnetic induction (*B*), which is the total flux of magnetic field lines through a unit cross sectional area of the material, considering both lines of force from the applied field and from the magnetisation of the material. *B*, *H* and *M* are related by equation 1a in S.I. units and by equation 1b in cgs units.

$$B = \mu_o (H + M)$$

$$B = H + 4 p M$$

In equation 1a, the constant μ_o is the permeability of free space ($4p \times 10^{-7}$ Hm^{-1}), which is the ratio of *B*/*H* measured in a vacuum. In cgs units the permeability of free space is unity and so does not appear in equation 1b. The units of *B*, *H* and *M* for both S.I. and cgs systems. Note that in the cgs system $4pM$ is usually quoted as it has units of Gauss and is numerically equivalent to *B* and *H*.

Another equation to consider at this stage is that concerning the magnetic susceptibility (*c*), equation 2, this is the same for S.I. and cgs units. The magnetic susceptibility is a parameter that demonstrates the type of magnetic material and the strength of that type of magnetic effect.

$$\chi = \frac{M}{H}$$

Sometimes the mass susceptibility (c_m) is quoted and this has the units of m^3kg^{-1} and can be calculated by dividing the susceptibility of the material by the density.

Another parameter that demonstrates the type of magnetic material and the strength of that type of magnetic effect is the permeability (m) of a material, this is defined in equation 3 (the same for S.I. and cgs units).

$$\mu = \frac{B}{H}$$

In the S.I. system of units, the permeability is related to the susceptibility, as shown in equation 4 and can also be broken down into m_0 and the relative permeability (m_r), as shown in equation 5.

$$mr = c + 1$$

$$m = mo\ mr$$

Finally, an important parameter (in S.I. units) to know is the magnetic polarisation (J), also referred to as the intensity of magnetisation (I). This value is effectively the magnetisation of a sample expressed in Tesla, and can be calculated as shown in equation.

$$J = moM$$

Table: The relationship between some magnetic parameters in cgs and S.I. units.

Quantity	Gaussian (cgs units)	S.I. Units	Conversion factor (cgs to S.I.)
Magnetic Induction (B)	G	T	10^{-4}
Applied Field (H)	Oe	Am^{-1}	$10^3 / 4\pi$
Magnetisation (M)	emu cm^{-3}	Am^{-1}	10^3
Magnetisation ($4\pi M$)	G	-	-
Magnetic Polarisation (J)	-	T	-
Specific Magnetisation (s)	emu g^{-1}	$JT^{-1}kg^{-1}$	1
Permeability (μ)	Dimensionless	H m^{-1}	$4\pi . 10^{-7}$
Relative Permeability(μ_r)	-	Dimensionless	-
Susceptibility (χ)	emu cm^{-3} Oe^{-1}	Dimensionless	4π

Maximum Energy Product (BH_{max})	M G Oe	k J m^{-3}	$10^2 / 4\pi$

(Where: G = Gauss, Oe = Oersted, T = Tesla)

CLASSIFICATION OF MAGNETIC MATERIALS

All materials can be classified in terms of their magnetic behaviour falling into one of five categories depending on their bulk magnetic susceptibility. The two most common types of magnetism are diamagnetism and paramagnetism, which account for the magnetic properties of most of the periodic table of elements at room temperature.

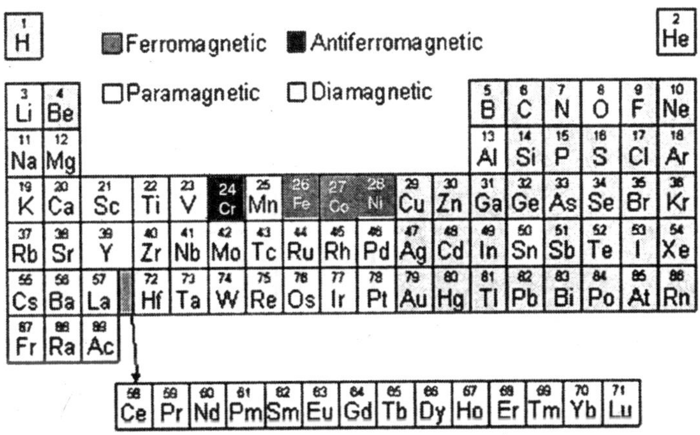

Fig. **A periodic table showing the type of magnetic behaviour of each element at room temperature.**

These elements are usually referred to as non-magnetic, whereas those which are referred to as magnetic are actually classified as ferromagnetic. The only other type of magnetism observed in pure elements at room temperature is antiferromagnetism. Finally, magnetic materials can also be classified as ferrimagnetic although this is not observed in any pure element but can only be found in compounds, such as the mixed oxides, known as ferrites, from which ferrimagnetism derives its name. The value of magnetic susceptible falls into a particular range for each type of material and this is shown in table 2 with some examples.

Table: Summary of different types of magnetic behaviour.

Type of Magnetism	Susceptibility	Atomic / Magnetic Behaviour		Example / Susceptibility	
Diamagnetism	Small & negative.	Atoms have no magnetic moment	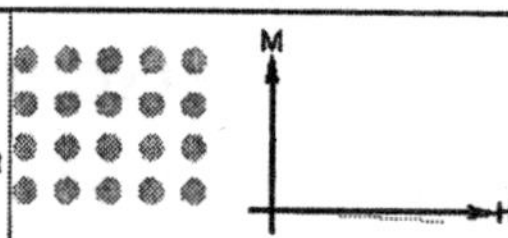	Au Cu	-2.74×10^{-6} -0.77×10^{-6}
Paramagnetism	Small & positive.	Atoms have randomly oriented magnetic moments	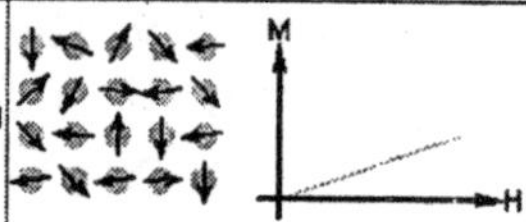	α-Sn Pt Mn	0.19×10^{-6} 21.04×10^{-6} 66.10×10^{-6}
Ferromagnetism	Large & positive, function of applied field, microstructure dependent.	Atoms have parallel aligned magnetic moments	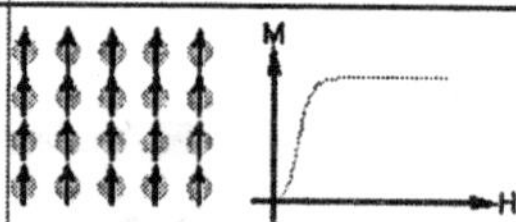	Fe	~100,000
Antiferromagnetism	Small & positive.	Atoms have mixed parallel and anti-parallel aligned magnetic moments	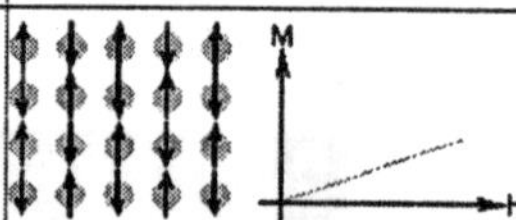	Cr	3.6×10^{-6}
Ferrimagnetism	Large & positive, function of applied field, microstructure dependent	Atoms have anti-parallel aligned magnetic moments	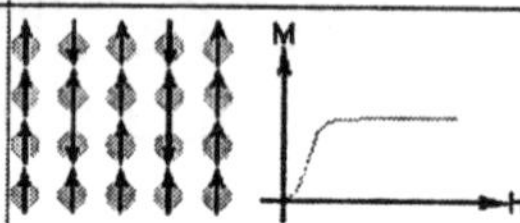	Ba ferrite	~3

Diamagnetism

In a diamagnetic material the atoms have no net magnetic moment when there is no applied field. Under the influence of an applied field (H) the spinning electrons precess and this motion, which is a type of electric current, produces a magnetisation (M) in the opposite direction to that of the applied field. All materials have a diamagnetic effect, however, it is often the case that the diamagnetic effect is masked by the larger paramagnetic or ferromagnetic term. The value of susceptibility is independent of temperature.

Paramagnetism

There are several theories of paramagnetism, which are valid for specific types of material. The Langevin model, which is true for materials with non-interacting localised electrons, states that each atom has a magnetic moment which is randomly oriented as a result of thermal agitation. The application of a magnetic field creates a slight alignment of these moments and hence a low magnetisation in the same direction as the applied field. As the temperature increases, then the thermal agitation will increase and it will become harder to align the atomic magnetic moments and hence the susceptibility will decrease. This behaviour is known as the Curie law and is shown in Equation, where C is a material constant called the Curie constant.

$$\chi = \frac{C}{T}$$

Materials which obey this law are materials in which the magnetic moments are localised at the atomic or ionic sites and where there is no interaction between neighbouring magnetic moments. The hydrated salts of the transition metals, e.g. $CuSO_4{\cdot}5H_2O$, are examples of this type of behaviour as the transition metal ions, which have a magnetic moment, are surrounded by a number of non-magnetic ions / atoms, which prevent interaction between neighbouring magnetic moments.

In fact the Curie law is a special case of the more general Curie-Weiss law, which incorporates a temperature constant (q) and derives from Weiss theory, proposed for ferromagnetic materials, that incorporates the interaction between magnetic moments.

$$\chi = \frac{C}{T - \theta}$$

In this equation q can either be positive, negative or zero. Clearly when q = 0 then the Curie-Weiss law equates to the Curie law. When q is non-zero then there is an interaction between neighbouring magnetic moments and the material is only paramagnetic above a certain transition temperature. If q is positive then the material is ferromagnetic below the transition temperature and the value of q corresponds to the transition temperature (Curie temperature, T_C). If q is negative then the material is antiferromagnetic below the transition temperature (Néel temperature, T_N), however the value of q does not relate to T_N. It is important to note that this equation is only valid when the material is in a paramagnetic state. It is also not valid for many metals as the electrons contributing to the magnetic moment are not localised. However, the law does apply to some metals, e.g. the rare-earths, where the 4f electrons, that create the magnetic moment, are closely bound.

The Pauli model of paramagnetism is true for materials where the electrons are free and interact to form a conduction band this is valid for most paramagnetic metals. In this model the conduction electrons are considered essentially to be free and under an applied field an imbalance between electrons with opposite spin is set up leading to a low magnetisation in the same direction as the applied field. The susceptibility is independent of temperature, although the electronic band structure may be affected, which will then have an effect on the susceptibility.

Ferromagnetism

Ferromagnetism is only possible when atoms are arranged in a lattice and the atomic magnetic moments can interact to align parallel to each other. This effect is explained in classical theory by the presence of a molecular field within the ferromagnetic material, which was first postulated by Weiss in 1907. This field is sufficient to magnetise the material to saturation. In quantum mechanics, the Heisenberg model of ferromagnetism describes the parallel alignment of magnetic moments in terms of an exchange interaction between neighbouring moments.

Weiss postulated the presence of magnetic domains within the material, which are regions where the atomic magnetic moments are aligned. The movement of these domains determines how the material responds to a magnetic field and as a consequence the susceptible is a function of applied magnetic field. Therefore, ferromagnetic materials are usually compared in terms of saturation magnetisation (magnetisation when all domains are aligned) rather than susceptibility.

In the periodic table of elements only Fe, Co and Ni are ferromagnetic at and above room temperature. As ferromagnetic materials are heated then the thermal agitation of the atoms means that the degree of alignment of the atomic magnetic moments decreases and hence the saturation magnetisation also decreases. Eventually the thermal agitation becomes so great that the material becomes paramagnetic; the temperature of this transition is the Curie temperature, T_C (Fe: T_C =770°C, Co: T_C =1131°C and Ni: T_C =358°C). Above T_C then the susceptibility varies according to the Curie-Weiss law.

Antiferromagnetism

In the periodic table the only element exhibiting antiferromagnetism at room temperature is chromium. Antiferromagnetic materials are very similar to ferromagnetic materials but the exchange interaction between neighbouring atoms leads to the anti-parallel alignment of the atomic magnetic moments. Therefore, the magnetic field cancels out and the material appears to behave in the same way as a paramagnetic material. Like ferromagnetic materials these materials become paramagnetic above a transition temperature, known as the Néel temperature, T_N. (Cr: T_N =37ºC).

Ferrimagnetism

Ferrimagnetism is only observed in compounds, which have more complex crystal structures than pure elements. Within these materials the exchange interactions lead to parallel alignment of atoms in some of the crystal sites and anti-parallel alignment of others. The material breaks down into magnetic domains, just like a ferromagnetic material and the magnetic behaviour is also very similar, although ferrimagnetic materials usually have lower saturation magnetisations. For example in Barium ferrite ($BaO.6Fe_2O_3$) the unit cell contains 64 ions of which the barium and oxygen ions have no magnetic moment, 16 Fe^{3+} ions have moments aligned parallel and 8 Fe^{3+} aligned anti-parallel giving a net magnetisation parallel to the applied field, but with a relatively low magnitude as only [! of the ions contribute to the magnetisation of the material.

INTRINSIC PROPERTIES OF MAGNETIC MATERIALS

The intrinsic properties of a magnetic material are those properties that are characteristic of the material and are unaffected by the microstructure (e.g. grain size, crystal orientation of grains). These properties include the Curie temperature, the saturation magnetisation and the magnetocrystalline anisotropy.

Saturation Magnetisation

The saturation magnetisation (M_S) is a measure of the maximum amount of field that can be generated by a material. It will depend on the strength of the dipole moments on the atoms that make up the material and how densely they are packed together. The atomic dipole moment will be affected by the nature of the atom and the overall electronic structure within the compound. The packing density of the atomic moments will be determined by the crystal structure (i.e. the spacing of the moments) and the presence of any non-magnetic elements within the structure.

For ferromagnetic materials, at finite temperatures, M_S will also depend on how well these moments are aligned, as thermal vibration of the atoms causes misalignment of the moments and a reduction in M_S. For ferrimagnetic materials not all of the moments align parallel, even at zero Kelvin and hence M_S will depend on the relative alignment of the moments as well as the temperature.

The saturation magnetisation is also referred to as the spontaneous magnetisation, although this term is usually used to describe the magnetisation within a single magnetic domain. Table 3 gives some examples of the saturation polarisation and Curie temperature of materials commonly used in magnetic applications.

Table: **The saturation polarisation (J_S) and Curie temperature (T_C) of a range of magnetic materials.**

Material (°C)	Magnetic Structure	J_S at 298K(T)	T_C
Fe	Ferro	2.15	770
Co	Ferro	1.76	1131

Ni	Ferro	0.60	358
$Nd_2Fe_{14}B$	Ferro	1.59	312
$SmCo_5$	Ferro	1.14	720
Sm_2Co_{17}	Ferro	1.25	820
$BaO.6Fe_2O_3$	Ferri	0.48	450
$SrO.6Fe_2O_3$	Ferri	0.48	450
Fe 3wtpercent Si	Ferro	2.00	740
Fe 4wtpercent Si	Ferro	1.97	690
Fe 35wtpercent Co	Ferro	2.45	970
Fe 78wtpercent Ni	Ferro	0.70	580
Fe 50wtpercent Ni	Ferro	1.55	500
$MnO.Fe_2O_3$	Ferri	0.51	300

Magnetic Anisotropy

In a crystalline magnetic material the magnetic properties will vary depending on the crystallographic direction in which the magnetic dipoles are aligned. Figure demonstrates this effect for a single crystal of cobalt. The hexagonal crystal structure of Co can be magnetised easily in the [0001] direction (i.e. along the c-axis), but has hard directions of magnetisation in the <1010> type directions, which lie in the basal plane (90° from the easy direction).

A measure of the magnetocrystalline anisotropy in the easy direction of magnetisation is the anisotropy field, *Ha* which is the field required to rotate all the moments by 90° as one unit in a saturated single crystal. The anisotropy is caused by a coupling of the electron orbitals to the lattice, and in the easy direction of magnetisation this coupling is such that these orbitals are in the lowest energy state.

The easy direction of magnetisation for a permanent magnet, based on ferrite or the rare earth alloys, must be uniaxial, however, it is also possible to have materials with multiple easy axes or where the easy direction can lie anywhere on a certain plane or on the surface of a cone. The fact that a permanent magnet has uniaxial anisotropy means that it is difficult to demagnetise as it is resistant to rotation of the direction of magnetisation.

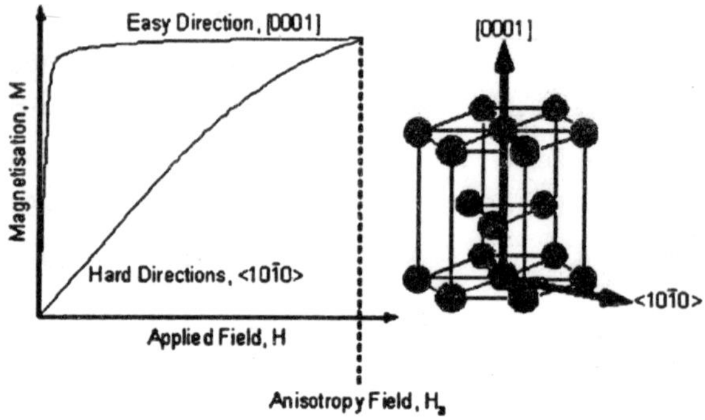

Fig. The Magnetocrystalline Anisotropy of Cobalt.

MAGNETIC DOMAINS

In order to explain the fact that ferromagnetic materials with spontaneous magnetisation could exist in the demagnetised state Weiss proposed the concept of magnetic domains. Weiss built on earlier work carried out by Ampère, Weber and Ewing suggesting their existence. The findings of this work revealed that within a domain large numbers of atomic moments are aligned typically 10^{12}-10^{18}, over a much larger volume than was previously suspected. The magnetisation within the domain is saturated and will always lie in the easy direction of magnetisation when there is no externally applied field. The direction of the domain alignment across a large volume of material is more or less random and hence the magnetisation of a specimen can be zero.

Magnetic domains exist in order to reduce the energy of the system. A uniformly magnetised specimen has a large magnetostatic energy associated with it. This is the result of the presence of magnetic free poles at the surface of the specimen generating a demagnetising field, H_d. From the convention adopted for the definition of the magnetic moment for a magnetic dipole the magnetisation within the specimen points from the south pole to the north pole, while the direction of the magnetic field points from north to south. Therefore, the demagnetising field is in opposition to the magnetisation of the specimen. The magnitude of H_d is dependent on the geometry and magnetisation of the specimen. In general if the sample has a high length to diameter ratio (and is magnetised in the long axis) then the demagnetising field and the magnetostatic energy will be low.

The break up of the magnetisation into two domains reduces the magnetostatic energy by half. In fact if the magnet breaks down into N domains then the magnetostatic energy is reduced by a factor of 1/N, has a quarter of the magnetostatic energy of figure. Figure shows a closure domain structure where the magnetostatic energy is zero, however, this is only possible for materials that do not have a strong uniaxial anisotropy, and the neighbouring domains do not have to be at 180º to each other.

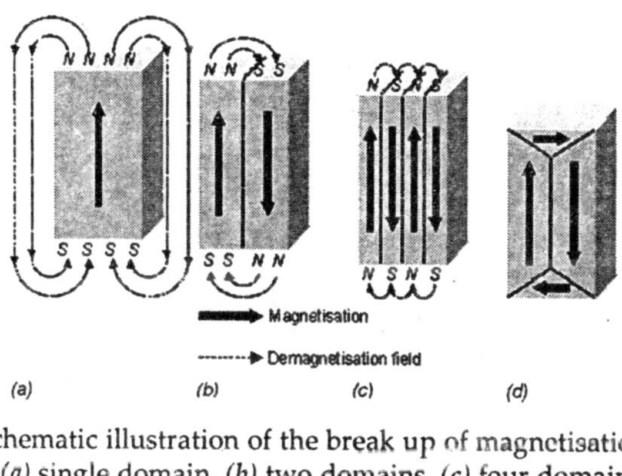

Fig. Schematic illustration of the break up of magnetisation into domains *(a)* single domain, *(b)* two domains, *(c)* four domains and *(d)* closure domains.

The introduction of a domain raises the overall energy of the system, therefore the division into domains only continues while the reduction in magnetostatic energy is greater than the energy required to form the domain wall. The energy associated a domain wall is proportional to its area. The schematic representation of the domain wall illustrates that the dipole moments of the atoms within the wall are not pointing in the easy direction of magnetisation and hence are in a higher energy state. In addition, the atomic dipoles within the wall are not at 180º to each other and so the exchange energy is also raised within the wall. Therefore, the domain wall energy is an intrinsic property of a material depending on the degree of magnetocrystalline anisotropy and the strength of the exchange interaction between neighbouring atoms. The thickness of the wall will also vary in relation to these parameters, as a strong magnetocrystalline anisotropy will favour a narrow wall, whereas a strong exchange interaction will favour a wider wall.

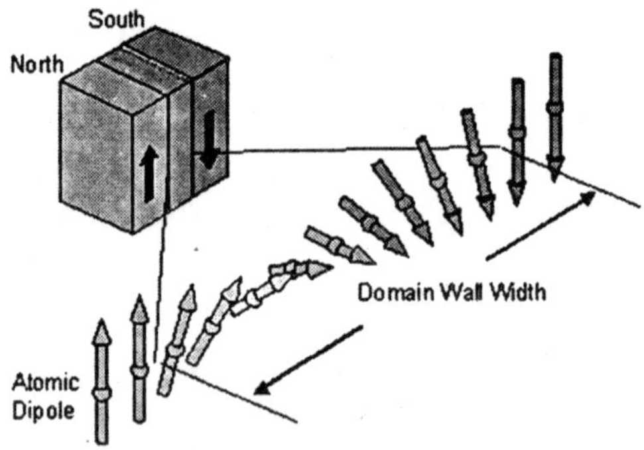

Fig. Schematic representation of a 180º domain wall.

A minimum energy can therefore be achieved with a specific number of domains within a specimen. This number of domains will depend on the size and shape of the sample (which will affect the magnetostatic energy) and the intrinsic magnetic properties of the material (which will affect the magnetostatic energy and the domain wall energy).

MAGNETIC HYSTERESIS

Ferromagnetic and ferrimagnetic materials have non-linear initial magnetisation curves as the changing magnetisation with applied field is due to a change in the magnetic domain structure. These materials also show hysteresis and the magnetisation does not return to zero after the application of a magnetic field. A typical hysteresis loop; the two loops represent the same data, however, the red loop is the polarisation ($J = m_0 M = B \quad m_0 H$) and the blue loop the induction, both plotted against the applied field.

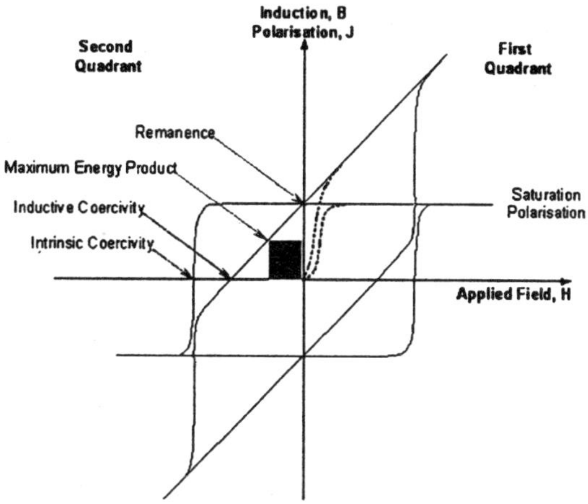

Fig. A typical Hysteresis loop for a Ferro- or Ferri- magnetic Material.

Illustrated in the first quadrant of the loop is the initial magnetisation curve (dotted line), which shows the increase in polarisation (and induction) on the application of a field to an unmagnetised sample. In the first quadrant the polarisation and applied field are both positive, i.e. they are in the same direction. The polarisation increases initially by the growth of favourably oriented domains, which will be magnetised in the easy direction of the crystal. When the polarisation can increase no further by the growth of domains, the direction of magnetisation of the domains then rotates away from the easy axis to align with the field. When all of the domains have fully aligned with the applied field saturation is reached and the polarisation can increase no further.

If the field is removed the polarisation returns along the solid red line to the y-axis (i.e. H=0), and the domains will return to their easy direction of magnetisation, resulting in a decrease in polarisation. In figure, the line from the saturation point to the y-axis is horizontal, which is representative of a well aligned material, where the domains are magnetised in the easy direction of the crystal at the saturation point.

If the direction of applied field is reversed (i.e. to the negative direction) then the polarisation will follow the red line into the second quadrant. The hysteresis means that the polarisation lags behind the applied field and will not immediately switch direction into the third quadrant (i.e. negative polarisation). The polarisation will only decrease after a sufficiently high field is applied to:

1. Nucleate and grow domains favourably oriented with respect to the applied field or
2. Rotate the direction of magnetisation of the domains towards the applied field

After applying a high enough field saturation polarisation will be achieved in the negative direction. If the applied field is then decreased and again applied in the positive direction then the full hysteresis loop is plotted.

If the field is repeatedly switched from positive to negative directions and is of sufficient magnitude then the polarisation and induction will cycle around the hysteresis loop in an anti-clockwise direction. The area contained within the loop indicates the amount of energy absorbed by the material during each cycle of the hysteresis loop.

Magnetic Parameters

The hysteresis loop is a means of characterising magnetic materials, and various parameters can be determined from it. From the first quadrant the saturation polarisation, J_S and hence the saturation magnetisation, M_S can be measured. Most of the useful information, however, can be derived from the second quadrant of the loop, and indeed it is conventional only to show this quadrant. The field that is produced by the magnet after the magnetising field has been removed is called the remanence, B_r or J_r. The reverse field required to bring the induction to zero is called the inductive coercivity, ${}_bH_c$, whereas the reverse field required to bring the magnetisation to zero is called the intrinsic coercivity, ${}_jH_c$. The maximum value of the product of B and H is called the maximum energy product, $(BH)_{MAX}$ and is a measure of the maximum amount of useful work that can be performed by the magnet. $(BH)_{MAX}$ is used as a figure of merit for permanent magnet materials.

In addition, the shape of the initial magnetisation curve and the hysteresis loop can provide information about the magnetic domain behaviour within the material. The *squareness factor* is a measure of how square the loop is and is a dimensionless quantity between 0 and 1, defined by the ratio of the reverse field required to reduce J by 10percent from the remanence to H_{cj}. A squareness factor of one therefore corresponds to a perfectly square loop. There are several other methods to quantify the squareness of the loop, such as the ratio of J_r to J_s.

Coercivity Mechanisms

There are various methods of increasing or decreasing the coercivity of magnetic materials, all of which involve the controlling the magnetic domains within the material. For a hard magnetic material it is desirable that the domains cannot easily rotate its direction of magnetisation and that the domain walls do not move easily and / or nucleation of domains is difficult.

To prevent easy rotation of domains the material could have a strong uniaxial magnetocrystalline anisotropy. Alternatively, shape anisotropy can occur in needle-like particles / grains, where the magnetostatic energy is less when the magnetisation is in the long axis of the needle compared to the short axis.

If the size of a magnetic particle / grain decreases then there is a critical size below which the decrease in magnetostatic energy by splitting into two domains is less than the increase in energy due to the introduction of the domain wall. Particles that are below this critical size are known as "single domain particles", and if they have sufficiently high anisotropy to prevent the easy rotation of the direction of magnetisation then the particles will be permanently magnetic and difficult to demagnetise. This type of coercivity mechanism can be observed in melt-spun NdFeB magnets where the crystal size is ~50nm, compared to the critical size for single domain particles of ~300nm.

Permanent magnets can also achieve their resistance to demagnetisation by pinning of the domain walls. In $Sm_2(Co,Fe,Cu,Zr)_{17}$ type magnets this is achieved by the presence of a $SmCo_5$ based phase in which the domain wall energy is significantly lower than that of the majority Sm_2Co_{17} based phase. The domain walls are therefore pinned within the $SmCo_5$ phase and both magnetisation and demagnetisation processes are difficult.

Permanent magnets can also achieve coercivity by making the nucleation of new domains difficult. This mechanism can be found in sintered NdFeB permanent magnets where a non-magnetic grain boundary phase acts to smooth the grain boundaries, removing domain nucleation sites. Nucleation controlled permanent magnets are easily magnetised as the initial state has several domains in each crystal, but are difficult to demagnetise because this would require the nucleation of new reverse domains

OBSERVATION OF MAGNETIC DOMAINS

There are many methods of imaging magnetic domains either directly or indirectly. Indirect techniques such as the Bitter method involve decorating the domain walls and imaging the decoration media. Direct methods however image the domain structure by the direct interaction of light, electrons or, more recently, a magnetic probe with the magnetic fields of the sample.

The Bitter Method

The Bitter method is a refinement of the well-known demonstration using iron fillings to map out the magnetic field in the vicinity of a bar magnet. In this technique a fine colloid suspension of Fe_3O_4 (magnetite), usually referred to as ferrofluid, is dropped on the surface of the specimen and a glass cover slide is placed over the top. Under the action of the gradient in the stray magnetic fields, the colloid particles are attracted to the regions of maximum magnetic field, which coincide with the lines at which the domain boundaries intersect the surface.

Magneto-optic Methods

These methods make use of the effect of the domain magnetization on the polarization of plane polarized light observed by reflection from the surface of a specimen (Kerr effect), or by transmission through the specimen (Faraday effect). Both of these techniques have the advantage that changes in the domain pattern can be followed and that the sample can be subjected to a magnetic field, heat, strain etc during the observation. The small order of the optical rotation, however can lead to experimental difficulties.

Kerr Effect

When plane polarized light illuminates the surface of a magnetic specimen, the reflected light is in general elliptically polarized. Figure shows a Kerr effect image from a sample of $Nd_{16}Fe_{76}B_8$ (atomicpercent). The image has been taken using a conventional metallographic microscope fitted with a polarizing filter, to plane polarize the incident beam, and an analyzing filter, in the reflected beam at ~90° to the polarizing filter, to generate magnetic domain contrast. The domain pattern shown is usually referred to as a "cog wheel" pattern and is observed when uniaxially anisotropic samples are sectioned perpendicular to their easy direction of magnetisation. When the sample is sectioned with the easy direction of magnetisation in the plane of the section then a striped pattern is observed, with the easy direction parallel to the stripes.

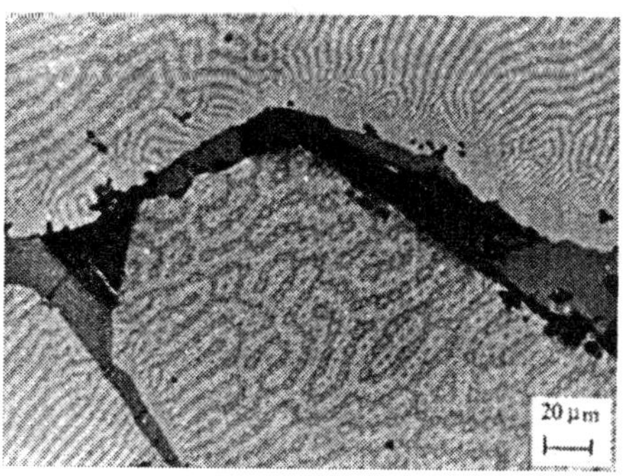

Fig. Kerr effect image from Homogenised $Nd_{16}Fe_{76}B_8$.

The Kerr effect has been developed as an accurate method of characterising the magnetic properties of thin film samples that could not be measured by conventional methods due to the small volume of magnetic material present. It has also been used as a method of recording digital data where ones and zeros are stored as magnetic domains magnetised in one direction or the other. The data is then read back using a laser spot scanned across the magnetic surface.

Faraday Effect

This is similar to the Kerr effect except that the rotation of the axis of polarization is caused during transmission of light through a ferromagnetic solid. Its use is therefore restricted to thin slices of ferromagnetic oxide or films and is thus less useful than the Kerr effect for domain observation.

HARD MAGNETIC MATERIALS

Hard magnets, also referred to as permanent magnets, are magnetic materials that retain their magnetism after being magnetised. Practically, this means materials that have an intrinsic coercivity of greater than ~10kAm^{-1}.

It is believed that permanent magnets have been used for compasses by the Chinese since ~2500BC. However, it was only in the early twentieth century that high carbon steels and then tungsten / chromium containing steels replaced lodestone as the best available permanent magnet material. These magnets were permanent magnets due to the pinning of domain walls by dislocations and inclusions. The movement of dislocations within a material is often hindered by the same factors that effect the motion of domain walls and as a consequence these steels are mechanically very hard and are the origin of the term hard magnetic. These magnets had an energy product of approximately 8kJm^{-3}.

Applications of Hard Magnetic Materials

Alnicos

In the development of permanent magnets the first improvement over steels came about in the early 1930s with the discovery of the group of alloys called the Alnico alloys.

These alloys are based mainly on the elements nickel, cobalt and iron with smaller amounts of aluminium, copper and titanium (Typical weightpercent: Fe-35, Co-35, Ni-15, Al-7, Cu-4, Ti-4). The alloy composition and processing were developed over the years and the properties reached a maximum in 1956 with the introduction of anisotropic columnar alnico 9, with an energy product of ~80kJm^{-3}. These alloys are still used today as they have a high Curie temperature (~850°C), and as a result can operate at higher temperatures as well as having more stable properties around room temperature than some of the more modern alloys. However, their main disadvantage is that they have low intrinsic coercivity (~50kAm^{-1}) and as a consequence must be made in the form of horseshoes or long thin cylinders, which cannot be exposed to significant demagnetising fields. The magnets are either sintered or directionally cast, and then annealed in a magnetic field. This processing route develops an oriented microstructure consisting of rods of strongly magnetic Fe-Co (a') in a matrix of weakly magnetic Ni-Al (a). The coercivity derives from the rod shaped nature of the a' phase generating shape anisotropy along with the weak magnetism of the a phase pinning the domain walls.

Hard Ferrites

The next advance in the development of permanent magnets came in the 1950s with the introduction of hard hexagonal ferrites, often referred to as ceramic magnets. These materials are ferrimagnetic and considering the proportion of iron within the material have quite a low remanence (~400mT). The coercivity of these magnets (~250kAm^{-1}), however, is far in excess of any previous material. The low remanence means that the maximum energy product is only ~40kJm^{-3}, which is lower than the alnicos, but due to the high coercivity these magnets can be made into thinner sections. The magnets could also be exposed to moderate demagnetising fields and hence could be used for applications such as permanent magnet motors. The hexagonal ferrite structure is found in both $BaO.6Fe_2O_3$ and $SrO.6Fe_2O_3$, but Sr ferrite has slightly superior magnetic properties.

The main advantage of ferrites is that they are extremely low cost, due to the ease of processing and the low cost of raw materials, which makes them the most widely used permanent magnet material. The magnets are made by a powder metallurgy processing route and there are no problems with oxidation of the powder during processing, as the material is already a stable oxide. The powder processing route ensures that the magnets comprise of very small grains (<1mm), which is essential for generating coercivity in these magnets. During processing the powder is compacted in a magnetic field in order to align the easy direction of magnetisation of the particles and hence enhances the remanence and the maximum energy product.

SmCo Type

In 1966 the magnetic properties of the YCo_5 phase were discovered. This was the first phase based on a rare-earth (RE) and a transition metal (TM) to be found to have permanent magnetic properties. The combination of RE and TM is ideal as the RE provides the anisotropy to the phase and the TM provides the high magnetisation and Curie temperature. The discovery of $SmCo_5$ soon followed in 1967 and this became the first commercial RE/TM permanent magnetic material, which was polymer bonded and had an energy product of ~40kJm^{-3}. It was later (1969) found that $SmCo_5$ sintered magnets could be made with energy products of the ~160kJm^{-3}. These magnets have excess Sm which forms a smoothing grain boundary phase and coercivity is achieved by prevention of the nucleation of reverse domains.

In 1976 the record maximum energy product was increased to 240kJm^{-3}, with a Sm_2Co_{17} based alloy. These materials are based on the general composition $Sm_2(Co,Fe,Cu,Zr)_{17}$ and achieve their permanent magnetic properties by careful control of the microstructure. The magnets are produced by powder metallurgy and are solution treated at ~1100ºC, where they are single phase. This homogenising stage is followed by several aging treatments at lower temperature where a cellular microstructure is formed. The cells are based on the Sm_2Co_{17} type phase, which is enriched in Fe and the cell boundaries comprise of a layer of $SmCo_5$ type phase, which is enriched in Cu. The intrinsic magnetic properties of the cells and the cell boundaries vary such that the magnetic domain wall energy is greatly reduced within the cell boundary and hence pin the domain walls, leading to permanent magnetic properties. There is still a great deal of interest in these materials as they have the potential for operating at high temperature (~500ºC), making new applications possible, for example as bearings in gas turbine engines.

The main problem with Sm/Co based magnets is the expense of the raw materials. Samarium is much less abundant than other light rare-earth elements, such as La, Ce, Pr and Nd, which account for over 90percent of rare-earth metals in typical rare-earth ores. Cobalt is classified as a strategically important metal and hence sales are restricted.

NdFeB Type

In 1984 the magnetic properties of NdFeB were discovered simultaneously by General Motors in the USA, and Sumitomo Special Metals of Japan. Both groups produced materials based on the magnetic phase $Nd_2Fe_{14}B$, but employed different processing routes.

SINTERING

In Japan, Sumitomo special metals developed a powder metallurgy processing route, which initially gave the highest ever observed energy product, in excess of $300kJm^{-3}$. Subsequently NdFeB based sintered permanent magnets have been produced with maximum energy products of $\sim 450kJm^{-3}$, by improved heat treatment, more controlled processing and the use of more iron rich compositions.

Sintered NdFeB based magnets achieve their coercivity by virtue of a Nd-rich phase at the grain boundaries which acts to produce liquid phase sintering, smooth the boundaries and hence prevent nucleation of reverse magnetic domains.

The processing route for sintered NdFeB based magnets is shown in figure. The as-cast ingot must first be broken into a powder. This is achieved most conveniently by exposing the ingot to hydrogen, which is absorbed at the surface. The hydrogen enters the material in the spaces between the atoms and causes the material to expand. The differential expansion generates stress in the ingot and the alloy breaks down into a fine powder. This process is known as Hydrogen Decrepitation (HD). The HD powder is then broken up further by a jet milling stage which reduces the particle size to around 5mm. It must be noted that when the alloy is in a powdered form then it is very flammable and must be handled under an inert gas.

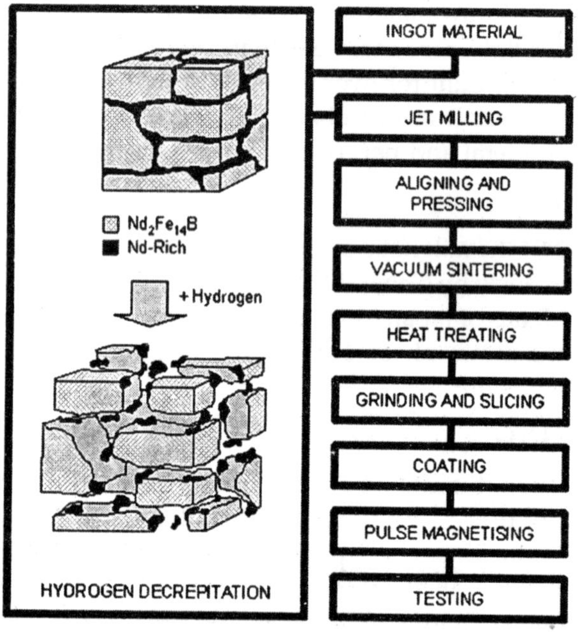

Fig. The Processing Route for Sintered NdFeB Permanent Magnets.

When the powder has been broken down to so fine a size, each particle of powder is a single crystal, which can be aligned in a magnetic field. This alignment is held in place by pressing the powder into a green compact, which is about 60percent dense. The compact is then heated in vacuum to ~1060ºC for 1 hour. During the heating stage the hydrogen comes out of the material and is pumped away. When held at ~1060ºC for 1 hour sintering occurs and the compact densifies, with the assistance of a liquid formed by the melting of the Nd-rich phase. After sintering the magnets are quenched and then heat treated in order to achieve the optimum magnetic properties.

The magnet must then be machined to the final dimensions required for the intended application. Due to the large degree of shrinkage that occurs during sintering, which is greater in the direction of alignment, it is not possible to press compacts that will shrink to the exact size required. The machining is a very expensive operation and, particularly for small magnets, a large proportion of the material may need to be machined away.

Due to the highly reactive nature of the Nd-rich phase, the magnets tend to corrode very rapidly, particularly in moist environments. Therefore, the next stage in the processing is to provide a protective barrier on the surface of the magnets. This is usually done with a nickel coating, although aluminium, zinc and epoxy resins are also used. Finally, the magnets are magnetised and tested prior to shipping to the customer.

MELT SPINNING

In the USA, melt-spinning was used to produce a ribbon like powdered material. In this process, molten alloy is ejected onto the surface of a rotating water cooled wheel, and cooling rates of the order of one million°C/s are achieved. The microstructure and magnetic properties of the NdFeB ribbons formed by melt-spinning are very sensitive to the quench rate, i.e. the speed of the rotating wheel. High quench rates produce essentially amorphous ribbons (i.e. no crystal grains) having negligible intrinsic coercivity. Optimum quench rates yield ribbons with the highest coercivities; they are comprised of roughly spherical $Nd_2Fe_{14}B$ grains (20-100 nm in diameter), which are single domain particles thus have a high coercivity (~1000kAm^{-1}). At wheel velocities below the optimum, the slow cooling rate produces ribbons that consist of larger grains and are characterised by low coercivities. This powder cannot be sintered to produce fully dense magnets without destroying the magnetic properties, but can be employed in one of three ways:

MQ-I

The melt spun ribbon is blended with a resin to produce a bonded permanent magnet. The crystals of MQI material are randomly oriented so that the magnets are isotropic and can be magnetised along any axis. The isotropic nature of the material limits the (BH)max to ~80 kJm^{-3}.

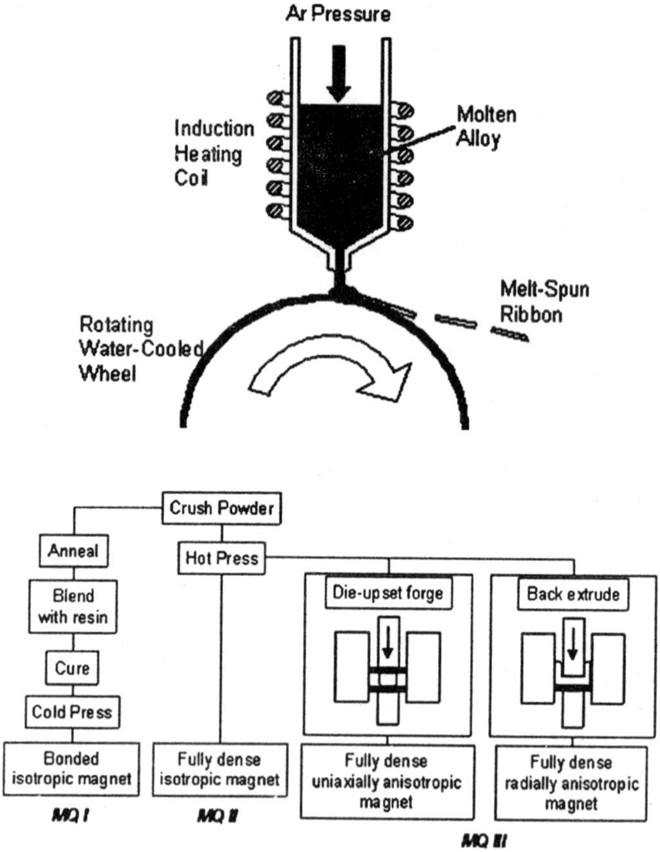

Fig. Schematic representation of the melt-spinning process and MQ magnet production.

MQ-II

Improved densification of melt-spun ribbons can be achieved by hot pressing at ~750ºC, without adversely effecting the coercivity of the powder. These "MQII" type magnets exhibit a slight (~10percent) degree of magnetic alignment and are 100percent dense, i.e. the magnetic properties are not diluted by a non-magnetic material, such as a resin. This gives MQII a higher (BH)max than MQI of 100-120 kJm^{-3}.

MQ III

Substantially greater alignment (> 75percent), and hence greater maximum energy products, can be obtained by an additional processing step. The MQII material is heated to ~750ºC in a die cavity having a larger diameter and then slowly deformed. This second hot press, termed die upset forging, produces plastic flow and a reorientation of the crystals. Such magnets, known as 'MQIII' are 100percent dense, and because of the alignment of crystals have maximum energy products of ~400 kJm^{-3}.

Applications of Hard Magnetic Materials

The rate of growth in the production of RE-Magnets has continued unabated, despite fluctuations in the world economy. Thus, sintered NdFeB magnets are exhibiting a current growth rate of ~12percent whereas the growth rate for bonded NdFeB magnets is in excess of 20percent. The total value of hard magnets now exceeds that of soft magnets and the gap is widening. The reason for this spectacular growth has been due partially to the booming global PC market, as around 60percent of NdFeB magnet production goes into disc-drive applications, primarily voice-coil-motors (VCMs). This is by no means the whole story however and Table 4 gives a summary of the very wide range of applications for RE-magnets, with many of these being capable of substantial further growth.

In general terms, permanent magnets are far more important than is generally realised and this is perhaps, best illustrated by their use in the motor car. In the early fifties a car would have one magnet (the speedometer) whereas some modern cars can have over a hundred permanent magnet motors. Currently these are almost exclusively based on Sr-ferrite ($SrFe_{12}0_{19}$) and the penetration of NdFeB magnets into this area requires a significant cost reduction, an increase in the maximum operating temperature and improvement in corrosion resistance. The potential benefits of using NdFeB magnets would be a significant reduction in volume and weight and an improved efficiency. This will probably be a major influence in the use of these magnets in the future.

Growing concern about global warming has scientific and technological implications and many of these impinge on the use of NdFeB magnets. Future uses could include their more widespread use in "white goods" such as washing machines, refrigerators etc, in order to improve energy efficiency and hence reduce CO_2 emissions. Another large use could be in generators for domestic combined heat and power units and in clean energy production such as windmills. The biggest potential however is in electric vehicles (EVs) which could be hybrid vehicles or totally driven by electricity in the form of batteries or a fuel cell. There has been an enormous increase of interest and activity in this area over the past 5 years and the Japanese have been the first to commercialise these vehicles.

Examples of applications for permanent magnetic materials

Automotive: Starter motors, Anti-lock braking systems (ABS), Motor drives for wipers, Injection pumps, Fans and controls for windows, seats etc, Loudspeakers, Eddy current brakes, Alternators.

Telecommunications: Loudspeakers, Microphones, Telephone ringers, Electro-acoustic pick-ups, Switches and relays.

Data Processing: Disc drives and actuators, Stepping motors, Printers.

Consumer Electronics: DC motors for showers, Washing machines, Drills, Low voltage DC drives for cordless appliances, Loudspeakers for TV and Audio, TV beam correction and focusing device, Compact-disc drives, Home computers, Video Recorders, Clocks.

Electronic and Instrumentation: Sensors, Contactless switches, NMR spectrometer, Energy meter disc, Electro-mechanical transducers, Crossed field tubes, Flux-transfer trip device, Dampers.

Industrial: DC motors for magnetic tools, Robotics, Magnetic separators for extracting metals and ores, Magnetic bearings, Servo-motor drives, Lifting apparatus, Brakes and clutches, Meters and measuring equipment.

Astro and Aerospace: Frictionless bearings, Stepping motors, Couplings, Instrumentation, Travelling wave tubes, Auto-compass.

Biosurgical: Dentures, Orthodontics, Orthopaedics, Wound closures, Stomach seals, Repulsion collars, Ferromagnetic probes, Cancer cell separators, Magnetomotive artificial hearts, NMR / MRI body scanner.

SOFT MAGNETIC MATERIALS

Soft magnetic materials are those materials that are easily magnetised and demagnetised. They typically have intrinsic coercivity less than 1000 Am^{-1}. They are used primarily to enhance and/or channel the flux produced by an electric current. The main parameter, often used as a figure of merit for soft magnetic materials, is the relative permeability m_r, where $m_r = B/m_0H)$, which is a measure of how readily the material responds to the applied magnetic field. The other main parameters of interest are the coercivity, the saturation magnetisation and the electrical conductivity.

The types of applications for soft magnetic materials fall into two main categories: AC and DC. In DC applications the material is magnetised in order to perform an operation and then demagnetised at the conclusion of the operation, e.g. an electromagnet on a crane at a scrap yard will be switched on to attract the scrap steel and then switched off to drop the steel. In AC applications the material will be continuously cycled from being magnetised in one direction to the other, throughout the period of operation, e.g. a power supply transformer. A high permeability will be desirable for each type of application but the significance of the other properties varies.

For DC applications the main consideration for material selection is most likely to be the permeability. This would be the case, for example, in shielding applications where the flux must be channelled through the material. Where the material is used to generate a magnetic field or to create a force then the saturation magnetisation may also be significant.

For AC applications the important consideration is how much energy is lost in the system as the material is cycled around its hysteresis loop. The energy loss can originate from three different sources: 1. hysteresis loss, which is related to the area contained within the hysteresis loop; 2. eddy current loss, which is related to the generation of electric currents in the magnetic material and the associated resistive losses and 3. anomalous loss, which is related to the movement of domain walls within the material. Hysteresis losses can be reduced by the reduction of the intrinsic coercivity, with a consequent reduction in the area contained within the hysteresis loop. Eddy current losses can be reduced by decreasing the electrical conductivity of the material and by laminating the material, which has an influence on overall conductivity and is important because of skin effects at higher frequency. Finally, the anomalous losses can be reduced by having a completely homogeneous material, within which there will be no hindrance to the motion of domain walls.

Iron-Silicon Alloys

These alloys are used for transformer cores and are known as electrical steels. In the power industry electrical voltage is almost always AC and at low frequency, 50-60Hz. At these frequencies eddy currents are generated in the transformer core. Alloying the Fe with Si has a large marked effect on the electric resistivity of the material, with an increase of a factor of 4 for 3wtpercentSi. Silicon also has the benefit of reducing the magnetostriction (i.e. length change on magnetisation) and the magnetocrystalline anisotropy. In addition, the material is used in the form of laminations, typically 0.3 to 0.7mm thick. The addition of too much silicon makes the material extremely brittle and difficult to produce, giving a practical limitation of 4wtpercent to the amount of Si that can be added. Recently, a technique has been developed to produce laminations with >6wtpercent Si, by a $SiCl_4$ chemical vapour deposition treatment to enrich the laminations with Si after forming the laminations. Typically most electrical steels will contain between 3 and 4 wtpercent Si.

For transformer applications the flux lies predominantly in the length of the laminations and therefore it is desirable to enhance the permeability in this direction. This is achieved by various hot and cold rolling stages to produce textured sheets, known as grain-oriented silicon-steel, with the [001] direction in the length of the lamination. The <001> type crystal directions are the easy directions of magnetisation and hence the permeability is greater. Figure shows the anisotropy of Fe and illustrates the two types of texture that can be achieved, which are known as cube-on-edge and cube texture. Note that the cube texture has two <001> type directions in the plane of the sheet and provides an advantage if E-shaped laminations are to be cut from the sheet.

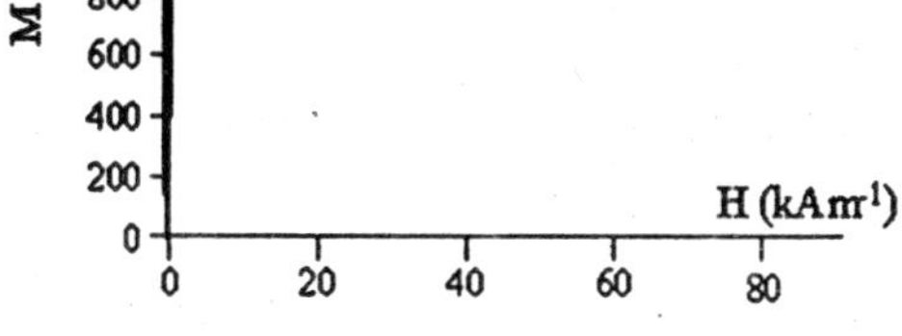

(a)

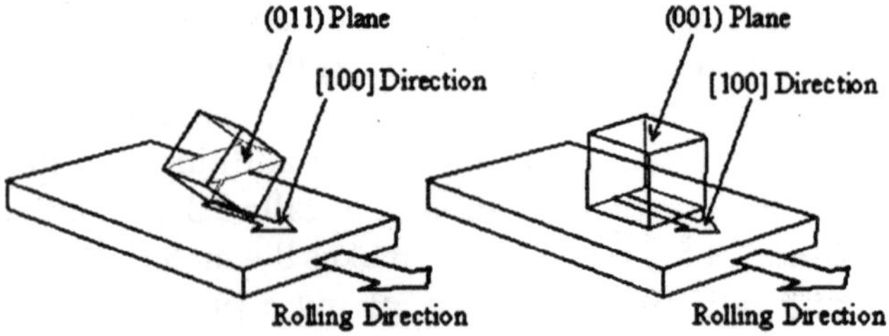

(b) *(c)*

Fig. (a) The magnetocrystalline anisotropy of Fe, (b) cube-on-edge texture and (c) cube texture in grain oriented silicon steel.

Amorphous & Nano-Crystalline Alloys

These materials can be produced in the form of a tape by melt-spinning. The alloys consist of iron, nickel and/or cobalt with one or more of the following elements: boron, carbon, phosphorous and silicon. They have extremely low coercivity, an order of magnitude less than standard Fe-Si, and consequently lower hysteresis losses. However, they have relatively low magnetisation and are not suitable for high current applications. They do find a market in low current applications and specialised small devices where they can compete with Ni-Fe.

Instead of casting the alloy onto a rotating wheel to produce tapes it is also possible to squirt a stream of molten alloy into a bath of water or oil to produce amorphous wires of typically 50mm thick. These wires show a very square hysteresis loop with large changes in magnetisation at low field, making them ideal for sensing and switching.

Recently there has been much interest in nano-crystalline material, which is produced by annealing the amorphous material. These alloys can be single phase but are usually comprised of nano-sized grains, in the range 10-50nm, in an amorphous matrix. They have relatively high resistivity, low anisotropy and good mechanical strength.

Nickel-iron Alloys

These alloys, known as permalloy, are extremely versatile and are used over a wide range of compositions, from 30 to

80wtpercentNi. Over this composition range the properties vary and the optimum composition must be selected for a particular application. The high Ni content alloys have high permeability; around 50wtpercentNi have high saturation magnetisation and low Ni content have a high electrical resistance.

There are special grades of Ni-Fe alloys that have zero magnetostriction and zero magnetic anisotropy, such as mumetal which is produced by a careful heat treatment and minor additions of Cu and Cr. These alloys have extremely high permeable, up to 300000 and intrinsic coercivity as low as $0.4Am^{-1}$.

Soft Ferrites

At high frequency metallic soft magnetic materials simply cannot be used due to the eddy current losses. Therefore, soft ferrites, which are ceramic insulators, become the most desirable material. These materials are ferrimagnetic with a cubic crystal structure and the general composition $MO.Fe_2O_3$, where M is a transition metal such as nickel, manganese or zinc.

MnZn ferrite, sold commercially as ferroxcube, can be used at frequencies up to 10MHz, for example in telephone signal transmitters and receivers and in switch mode power supplies (also referred to as DC-DC converters). For these type of application the driving force to increase frequency is to allow miniaturisation.

Additionally, part of the family of soft ferrites, are the microwave ferrites, e.g. yttrium iron garnet. These ferrites are used in the frequency range from 100MHz to 500GHz, for waveguides for electromagnetic radiation and in microwave devices such as phase shifters.

AC Applications

AC applications relate to electric circuits and primarily in transformers converting one AC voltage into another. Power transmission is more efficient at high voltage, but more dangerous and less easily used in the home. Therefore, step-up transformers are used to increase the voltage for transmission and step-down transformers are used to reduce the voltage before entering the home.

The smallest type of transformer is a DC-DC converter, also known as a switch mode power supply. These are often mounted on a chip that can be put onto a circuit board. They take a DC voltage input, oscillate the voltage to an AC signal, which then goes through a coil around a toroidal core, a pick-up coil picks up the signal from the core and rectifies it to the output voltage. The relative number of turns on the primary (input) and secondary (output) coils determines the difference in voltage between input and output.

Soft magnetic materials also play an important role in electric motors where they enhance the field produced by the motor windings. In permanent magnet motors they are also used to channel the flux produced by the permanent magnets.

DC Applications

One of the main DC applications is in the field of magnetic shielding. A high permeability magnetic material is used to encapsulate the device that requires shielding. Figure illustrates a simple example of magnetic shield where a tube/sphere of high permeability material channels the magnetic field away from the inside of the tube/sphere. The effectiveness of the shield can be expressed in terms of the shielding factor, S, where S relates to the field outside (B_o) and inside (B_i) the shield by equation.

$$S = \frac{B_0}{B_i}$$

For a sphere, *S* can be calculated by equation, where m_r = relative permeability, *d* = thickness of wall and *D* = diameter.

$$S = \frac{4}{3}\frac{\mu_r d}{D}$$

It is clear from equation that the higher the permeability then the better the material will be at shielding. It is also apparent that as the shield gets larger then either the permeability of the material or the wall thickness must increase to compensate.

Soft magnetic materials are also used for electromagnetic pole-pieces, to enhance the fields produced by the magnet. Solenoid switches also rely on soft magnetic materials to activate the switches. Most permanent magnet devices will use soft magnetic materials to channel flux lines or provide a return path for magnetic fields, e.g. MRI body scanners have large permanent magnets with a yoke of soft magnetic material to prevent self demagnetising fields that would reduce the field in the gap of the scanner.

At high frequency metallic soft magnetic materials simply cannot be used due to the eddy current losses. Therefore, soft ferrites, which are ceramic insulators, become the most desirable material. These materials are ferrimagnetic with a cubic crystal structure and the general composition $MO.Fe_2O_3$, where M is a transition metal such as nickel, manganese or zinc.

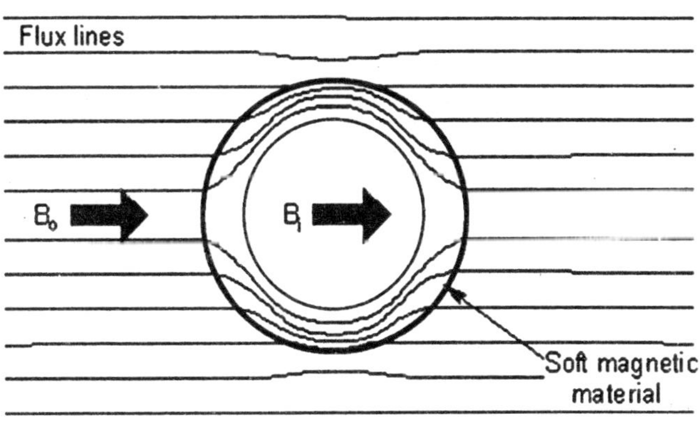

Fig. Magnetic shielding by a tube/sphere of soft magnetic material.

MnZn ferrite, sold commercially as ferroxcube, can be used at frequencies up to 10 MHz, for example in telephone signal transmitters and receivers and in switch mode power supplies (also referred to as DC-DC converters). For these type of application the driving force to increase frequency is to allow miniaturisation.

Additionally, part of the family of soft ferrites, are the microwave ferrites, e.g. yttrium iron garnet. These ferrites are used in the frequency range from 100MHz to 500GHz, for waveguides for electromagnetic radiation and in microwave devices such as phase shifters.

MAGNETIC RECORDING

Magnetic recording of the human voice was first carried out in 1898 by a Danish engineer, named Poulsen. The recording was made on a ferromagnetic wire, but due to a lack of amplification the recording quality was very poor. In 1927, magnetic tape was simultaneously invented in the USA and Germany. However, the familiar oxide tapes for audio recording were not developed until 1947, by the 3M corporation.

The first hard disk drive, known as the "RAMAC", was introduced by IBM in 1957, and had a storage capacity of 2000 bits in^{-2}. The storage capacity has rapidly and steadily increased since the "RAMAC" and in the year 2000 reached approximately 10 Gbits in^{-2}, i.e. an increase by a factor of five million. This increase in storage capacity is illustrated in figure, in terms of the decrease in area on the disk required to store the information.

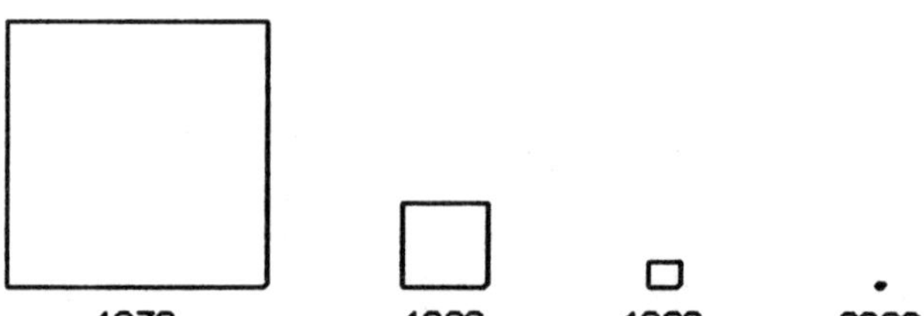

Fig. The block represents the size of one bit of data on a hard disk. The sizes have been enlarged 1000 times.

Magnetic Tapes

Magnetic tapes are extensively used for recording audio and video signals, although it is unclear how long this technology will continue to be used with the rising popularity of the digital versatile disk (DVD).

Tapes can be made with either a particulate media adhered to a plastic substrate or a metal evapourated (ME) film on the substrate. The magnetic layer on a particulate tape is only 40percent magnetic material whereas ME tapes have a 100percent magnetic layer. Therefore, ME tapes give better quality recording, but they are more time consuming to produce and are more expensive. Particulate tapes are much cheaper and hence account for the majority of magnetic tapes.

There is a range of magnetic particulates that can be used for tapes and these are listed in table 5, with their magnetic properties. The values of coercivity quoted in table 4 are approximate averages as the coercivity is highly dependent on particle size and shape, which will vary in any batch of powder. Smaller particle size and higher magnetisation lead to better quality recording, i.e. greater signal to noise ratio. As the density of recording increases then the magnetic domain size decreases and so the particle size must decrease and the coercivity of the material must increase. The cheapest and most commonly used material is g-Fe_2O_3, while increasing quality is achieved through the use of cobalt modified (2-3wtpercent) g-Fe_2O_3, chromium dioxide and, the best, pure iron. Finally, also mentioned in table 5, is barium hexaferrite, which has the highest coercivity and is used for applications where the stored data must be secure and is unlikely to require rewriting, e.g. a credit card.

Table: Approximate Magnetic Properties of Particulates used in Recording Media.

Material	Saturation Polarisation (mT)	Intrinsic Coercivity (kAm^{-1})	Average Particle Size (mm)	Particle Shape
g-Fe_2O_3	440	30	0.5 x 0.1	Needle
Co modified g-Fe_2O_3	460	60	0.5 x 0.1	Needle

CrO_2	600	70	0.4 x 0.05	Needle
Fe	2100	125	0.15 x 0.05	Needle
$BaO.6Fe_2O_3$	460	200	0.15 x 0.05	Disc

The production route for a particulate magnetic tape is illustrated in figure. The magnetic particles are mixed with a binder (dissolved in a solvent), lubricants that will help reduce the friction when then tape is moved over the head and abrasives (such as Al_2O_3) that are hard and help prevent wear of the magnetic tape. This mixture is poured onto a PET (polyethylene tetraphthalate) substrate, which is ~25mm thick. Sometimes aramid substrates are used for long play cassettes, as these substrates can be as thin as 5mm. The particles are magnetically anisotropic, usually due to their shape, and the next stage of the process is to align these particles in the length of the tape while the magnetic layer is still liquid. The solvent is then evapourated by heating the tape and it is rolled to improve the density and leave a magnetic layer of about 3-5mm thick.

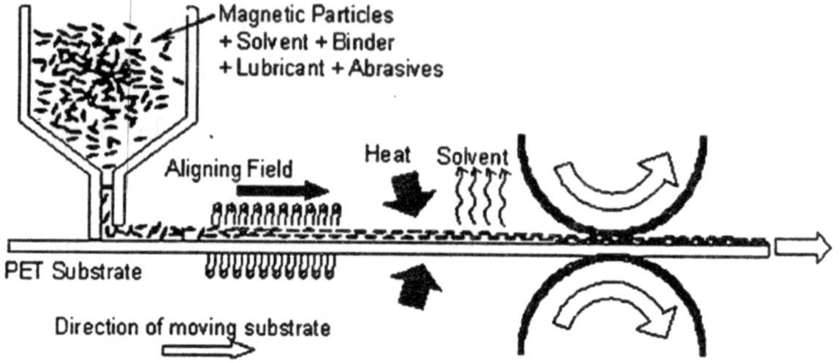

Fig. The processing route for particulate magnetic tape.

Magnetic Disks

Recording data onto a disk has obvious advantages with respect to access times, as the head can readily be moved to the appropriate place on the disk whereas a tape would need to be rewound or advanced. There are two types of disk: floppy and hard. The principles of manufacturing and recording on floppy disks are very similar to that of particulate magnetic tape, i.e. the same particulate materials on a plastic substrates.

Hard disk drives are formed on a rigid substrate, usually aluminium, which is around 2mm thick. On to the substrate are deposited several layers: an underlayer to help adhesion (~10nm nickel phosphide); a layer of chromium (5-10nm) to control orientation and grain size of magnetic layer; the magnetic layer (50nm PtCo with various additions of Ta, P, Ni, Cr); a protective overcoat (e.g. 10-20nm zirconia) and finally lubricant to reduce friction and wear of the disk (e.g. a monolayer of long chain fluorocarbons). The magnetic layer forms a cellular structure of Co-rich magnetic cells in a non-magnetic matrix. These cells act just like particulate recording media but on a much finer scale.

The construction of a hard disk drive is shown schematically in figure. The disk is attached to a spindle motor that will spin the disk; for greater storage capacity several disks can be built into a stack. Read and write heads are attached to a swinging arm (one for each side of the disks) that can be scanned across the disk using a voice coil.

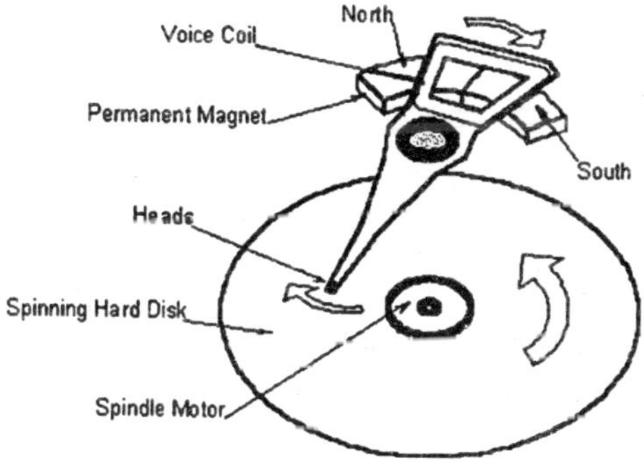

Fig. Schematic Representation of the Construction of a Hard Disk Drive.

Writing & Reading Data

Writing of data is performed using an inductive head, as illustrated in figure. Reading of data also uses an inductive head or, as in modern hard disk drives, a giant magnetoresistive (GMR) head.

The writing process involves passing a current (i.e. the signal to be recorded) through the coil of the head. This current generates a field in the air gap of the C-shaped core and a fringing field (in the plane of the tape or disk) that extends out of the gap to the tape or disk that is moving past it. The fringing field will change the magnetic state of the media and if the magnetic properties of the media are appropriate then the remanence of the tape in that region will be proportional to the amount of current applied to the coil. For digital signals only two remanent states are required for the material and hence the material requirements are not as stringent as for analogue recording, although smaller particle size is desired for high storage capacity and faster access time.

The reading process when carried out with an inductive head is very similar to the writing process. The magnetic fields extending out from the tape or disk induce a field in the C-shaped core of the read head, which in turn generates a voltage in the coil. This voltage can then be turned back into the original signal, be it audio, visual or digital data.

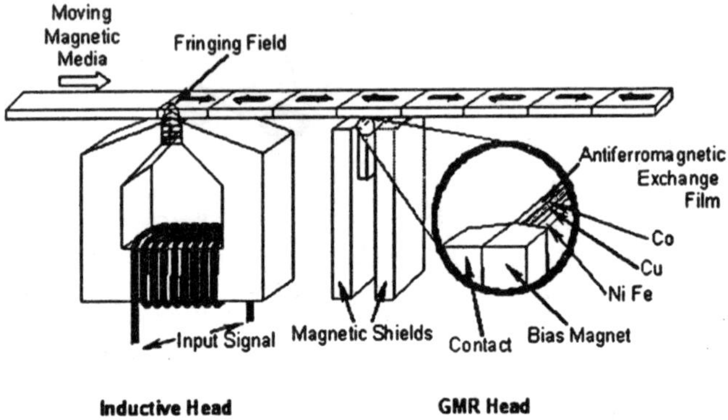

Fig. Schematic of an Inductive read / write head and a GMR read head.

In modern hard disk drives the data is read back using a GMR head. The GMR element of this head experiences changes in electrical resistance under the influence of a magnetic field. They can sense very low magnetic fields and have a very high spatial resolution. The use of GMR heads is essential as the increased storage density and smaller magnetic domain size in the media means that the field strength, at a specific height above the surface of the media, has decreased. The GMR element shown in figure is a spin valve type with 4 layers: an antiferromagnetic exchange film (e.g. iron/manganese); a layer of cobalt with its direction of magnetisation pinned by the antiferromagnetic exchange film (upwards in fig.17); a layer of copper which is a spacer and a layer of nickel/iron with its direction of magnetisation free to move under the influence of the magnetic field from the recording media. The biasing magnetic layer magnetises the NiFe in the plane of the film and perpendicular to the direction of magnetisation of the Co film. When the direction of magnetisation of the NiFe moves towards the direction of magnetisation of the Co then there is a drop in resistance; when it moves away then there is an increase in the resistance.

OTHER MAGNETIC MATERIALS

There are various other types of magnetic phenomena and materials that are exploited commercially. This section will look at magnetostrictive and magnetoresistive materials.

Magnetostriction

Most magnetic materials exhibit magnetostriction, which is a change in physical dimension as a result of magnetic ordering. There are two types of magnetostriction: spontaneous magnetostriction which arises from the magnetic ordering of atomic moments below the Curie temperature (usually causing a volume expansion) and field induced magnetostriction which arises from the alignment of magnetic domains on application of an applied field.

Spontaneous magnetostriction is exploited in the NiFe based alloy known as invar, which is noted for the fact that it shows zero thermal expansion up to its Curie temperature. As mentioned previously, the saturation magnetisation of a material decreases on heating towards the Curie temperature due to a decrease in the degree of alignment of the atomic magnetic moments. As this alignment and the saturation magnetisation decreases then so does the volume expansion caused by the spontaneous magnetostriction and hence the material contracts. In the case of invar this contraction due to loss of spontaneous magnetostriction is equal to the expansion caused by normal thermal vibration processes, and hence the material shows no change in dimensions. However, above the Curie temperature there is no longer any magnetic ordering and thermal expansion occurs normally.

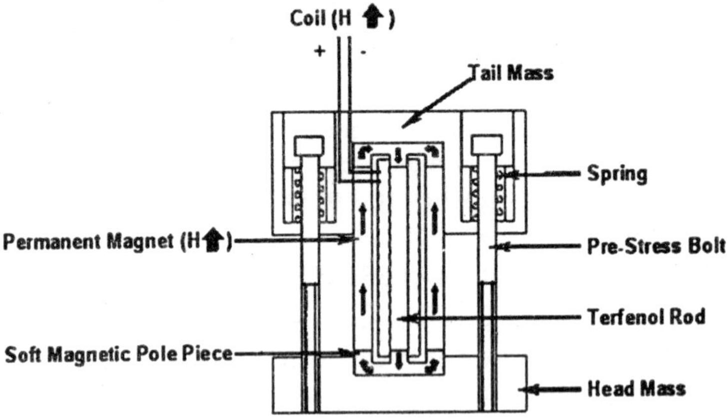

Fig. Schematic of a Terfenol Actuator

Field induced magnetostriction occurs when magnetic domains align and the dimension change caused by the ordering of the atomic magnetic moments all combine to give a dimensional change to the block of material. The material that shows the largest useful magnetostriction is Terfenol, which is an alloy of Tb, Dy and Fe. Terfenol is used for position and field sensors, as well as mechanical actuators and speakers. Magnetostrictive position / load sensors work by the fact that when a magnetostrictive material experiences a strain then the magnetisation of the material will change. Terfenol actuators usually have a rod of Terfenol which is placed under compression in order to align the magnetic domains perpendicular to the length of the rod. A coil around the rod can then apply a field to the rod that will align the domains along its length. The schematic in figure shows an actuator that has a permanent magnet to bias the Terfenol rod so that currents flowing in the coil in different directions will make the actuator decrease or increase in length. If the magnetic bias was not present then a current in either direction would cause the actuator to expand and if the actuator was being used as a loud speaker then the frequency of voltage applied to the coil would be doubled by the speaker.

Magnetoresistance

Magnetoresistance (MR) is the effect by which the electrical resistance of a magnetic material changes depending on the relative direction of the current and the magnetisation. In most cases the electrical resistance is highest when the current and magnetisation are parallel and lowest when they are perpendicular. The level of magnetoresistance shown by a material is usually expressed in terms of the percentage change in resistance from the highest to the lowest resistance and is usually of the order of a few percent. The main application for MR sensors is in the read heads of hard disk drives.

A similar effect has been observed in magnetic multilayers, e.g. Fe/Cr, where the alternate ferromagnetic layers couple anti-parallel to each other. Under the influence a magnetic field the relative orientation of the magnetisation of the layers changes and the electrical resistance decreases to a minimum when the magnetisation directions of the layers are parallel. This effect gives rise to resistance changes of 50-80percent, and the effect has been termed giant magnetoresistance (GMR).

Chapter 6

Electromagnetic Radiation

Electromagnetic radiation is a transverse energy wave that is composed of an oscillating electric field component, E, and an oscillating magnetic field component, M. The electric and magnetic fields are orthogonal to each other, and they are orthogonal to the direction of propogation of the wave. A wave is described by the wavelength, λ, which is the physical length of one complete oscillation, and the frequency v,which is the number of oscillations per second. The figure shows one wavelength of a wave of light. The names we give electromagnetic radiation for different wavelength and frequency ranges are listed in the electromagnetic spectrum document.

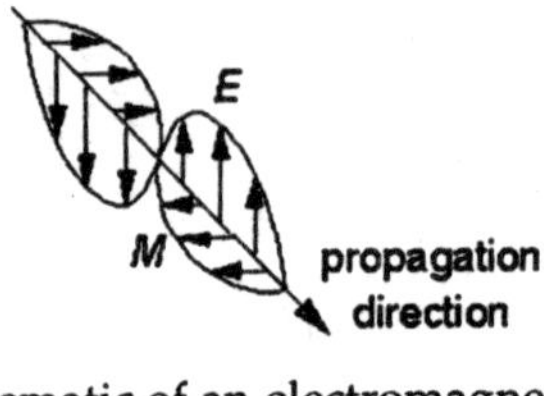

Fig. Schematic of an electromagnetic wave

VELOCITY OF LIGHT

Electromagnetic waves travel through a vacuum at a constant velocity of 2.99792×10^8 m/s, which is known as the speed of light, c. The relationship between the speed of light, wavelength, and frequency is:

$$c = \lambda\nu$$

When light passes through other media, the velocity of light decreases. For a given frequency of light, the wavelength also must decrease. This decrease in velocity is quantitated by the refractive index, n, which is the ratio of c to the velocity of light in another medium, v:

$$n = c / v$$

Since the velocity of light is lower in other media than in a vacuum, n is always a number greater than one. The table lists the refractive index of several examples. Refractive index is an intrinsic physical property of a substance, and can be used to monitor purity or the concentration of a solute in a solution. The refractive index of a material is measured with a refractometer, and is usually made versus air. If the precision warrants, the measurements can be corrected for vacuum. Note that the difference between n_{air} and n_{vacuum} is only significant in the fourth decimal place.

For anisotropic materials, such as quartz crystals, light of different polarizations will experience different refractive indices. These indices are called the ordinary refractive index, n_o, and the extraordinary refractive index, n_e.

medium	n*
air	1.0003
water	1.333
50percent sucrose in water	1.420
carbon disulfide	1.628
crystalline quartz	1.544 (n_o)
1.553 (n_e)	
diamond	2.417

*measured with 589.3 nm light

POLARIZATION

An incoherent light source, such as the hot filament of a light bulb, consists of multiple, randomly oriented light emitters, which produce electromagnetic waves with their electric-field vector oriented in all directions. The resulting light emission is called unpolarized light.

Linearly (or plane) polarized light is light in which the electric-field vector is oscillating in only one direction. Linearly polarized light is produced by isolating one orientation of the electric field with a polarizer, or from lasers that contain polarized optical components.

Circularly polarized light is light in which the electric field vector is rotating around the axis of light propogation. The electric field vector can rotate in either the right or left direction (as viewed in the direction of light propogation), and the light is called right circularly polarized or left circularly polarized, respectively.

WAVE-PARTICLE DUALITY

Electromagnetic radiation shows both wave and particle characteristics depending on how the radiation is observed. Einstein first postulated that the energy of radiation is quantized and that radiation is composed of energy packets that were later named photons. The energy, E, of one photon depends on its frequency (or wavelength):

$$E = h\nu = hc/\lambda$$

where h is Planck's constant (6.62618×10^{-34} Js), ν is the frequency of the radiation, c is the speed of light, and λ is wavelength.

DE BROGLIE EQUATION

Moving particles; such as electrons, protons, and neutrons; have wave properties as described by the de Broglie equation:

$$\lambda = h / p$$

where λ is wavelength, h is Planck's constant, and p is the momentum of the particle. Beams of particles can therefore show wave effects such as interference.

ELECTROMAGNETIC SPECTRUM

VISIBLE SPECTRUM

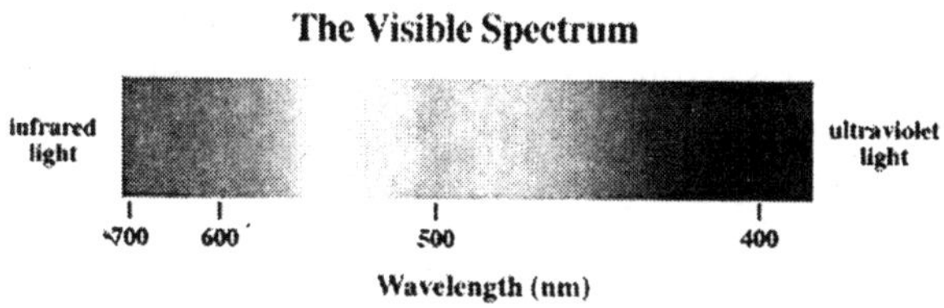

Figure. Electromagnetic Spectrum

Type of	Frequency	Wavelength	Type of Transition
Radiation	Range (Hz)	Range	
gamma-rays	10^{20}–10^{24}	$<10^{-12}$ m	Nuclear
x-rays	10^{17}–10^{20}	1 nm-1 pm	inner electron
ultraviolet	10^{15}–10^{17}	400 nm-1 nm	outer electron
visible	$4\text{-}7.5 \times 10^{14}$	750 nm-400 nm	outer electron
near-infrared	$1 \times 10^{14} - 4 \times 10^{14}$	2.5 um-750 nm	outer electron molecular vibrations
infrared	$10^{13} - 10^{14}$	25 um - 2.5 um	molecular vibrations
microwaves	3×10^{11}-10^{13}	1 mm - 25 um	molecular rotations,
electron			spin flips*
radio waves flips*	$<3 \times 10^{11}$	>1 mm	nuclear spin

*energy levels split by a magnetic field

SPECTROSCOPY

Electron spectroscopies analyse the electrons that are ejected from a material for qualitative or semi-quantitative analysis. In general an excitation source such as x-rays or electrons will eject an electron from an inner-shell orbital of an atom. Detecting photoelectrons that are ejected by x-rays is call x-ray photoelectron spectroscopy (XPS) or electron spectroscopy for chemical analysis (ESCA). Detecting electrons that are ejected from higher orbitals to conserve energy during electron transitions is called Auger electron spectroscopy (AES). These electron processes are described below. Ejected electrons can escape only from a depth of approximately 3 nm or less, making electron spectroscopy most useful to study surfaces of solid materials. Depth profiling is accomplished by combining an electron spectroscopy with a sputtering source that removes surface layers.

AUGER ELECTRON SPECTROSCOPY (AES)

Auger (pronounced ~o-jay) electron spectroscopy is an electron spectroscopic method that uses a beam of electrons to knock electrons out of inner-shell orbitals. Auger electrons are ejected to conserve energy when electrons in higher shells fill the vacancy in the inner shell. These Auger electrons have energies characteristic of the emitting atom due to the characteristic energy-level structure of that element.

Fig. Auger electron spectrometer

X-RAY PHOTOELECTRON SPECTROSCOPY (XPS, ESCA)

X-ray photoelectron spectroscopy (XPS, also called electron spectroscopy for chemical analysis, ESCA) is a electron spectroscopic method that uses x-rays to eject electrons from inner-shell orbitals. The kinetic energy, E_k, of these photoelectrons is determined by the energy of the x-ray radiation, h,v and the electron binding energy, E_b, as given by:

$E_k = h\nu - E_b$

The experimentally measured energies of the photoelectrons are given by:

$E_k = h\nu - E_b - E_w$

where E_w is the work function of the spectrometer.

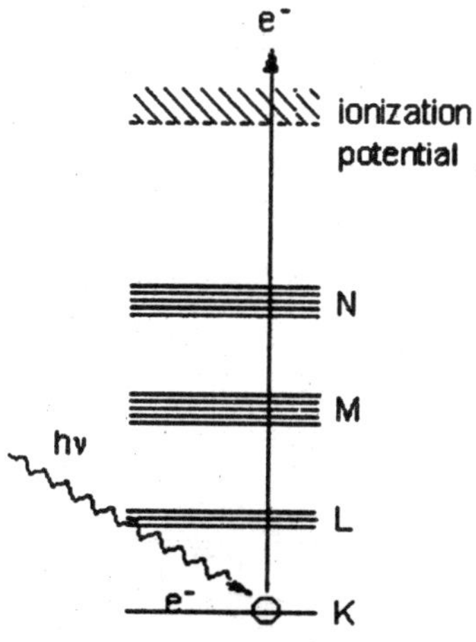

Fig. Energy-level diagram for XPS

The electron binding energies are dependent on the chemical environment of the atom, making XPS useful to identify the oxidation state and ligands of an atom.

XPS instruments consist of an x-ray source, an energy analyser for the photoelectrons, and an electron detector. The analysis and detection of photoelectrons requires that the sample be placed in a high-vacuum chamber. Since the photoelectron energy depends on x-ray energy, the excitation source must be monochromatic. The energy of the photoelectrons is analysed by an electrostatic analyser, and the photoelectrons are detected by an electron multiplier tube or a multichannel detector such as a microchannel plate.

INFRARED ABSORPTION SPECTROSCOPY (IR)

IR spectroscopy is the measurement of the wavelength and intensity of the absorption of mid-infrared light by a sample. Mid-infrared light (2.5 - 50 μm, 4000 - 200 cm^{-1}) is energetic enough to excite molecular vibrations to higher energy levels. The wavelength of IR absorption bands are characteristic of specific types of chemical bonds, and IR spectroscopy finds its greatest utility for identification of organic and organometallic molecules.

MECHANISM OF IR ABSORPTION

The transition moment for infrared absorption is:

$$R = < X_i \mid u \mid X_j \, dt >$$

where X_i and X_j are the initial and final states, respectively, and u is the electric dipole moment operator:

$$u = u_o + (r - r_e)(du/dr) + \ldots \text{ higher terms.}$$

u_o is the permanent dipole moment, which is a constant, and since $< X_i \mid X_j > = 0$, R simplifies to:

$$R = < X_i \mid (r - r_e)(du/dr) \mid X_j >$$

The result is that there must be a change in dipole moment during the vibration for a molecule to absorb infrared radiation.

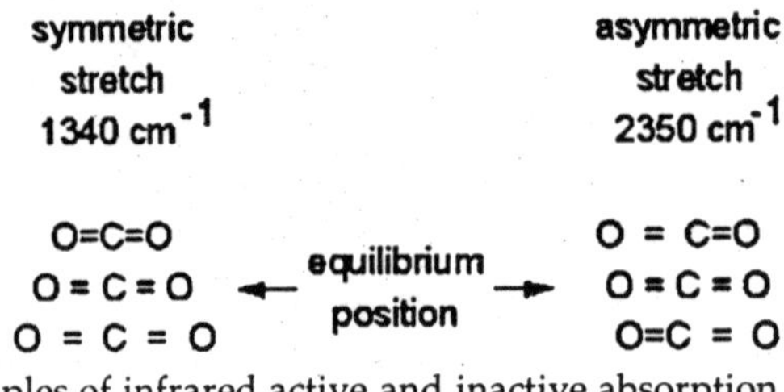

Figure. Examples of infrared active and inactive absorption bands in CO_2

There is no change in dipole moment during the symmetric stretch vibration and the 1340 cm^{-1} band is not observed in the infrared absorption spectrum (the symmetric stretch is called infrared inactive). There is a change in dipole moment during the asymmetric stretch and the 2350 cm^{-1} band does absorb infrared radiation (the asymmetric stretch in infrared active). A related vibrational spectroscopic method is Raman spectroscopy, which has a different mechanism and therefore provides complementary information to infrared absorption.

INFRARED ABSORPTION BANDS

IR absorption spectroscopy uses mid-infrared light (2.5 - 50 μm, 4000 – 200 cm^{-1}) to detect specific types of chemical bonds in a sample for identification of organic and organometallic molecules.

Table of characteristic IR bands

Group	Bond	Appox. Energy (cm^{-1})
hydroxyl	O-H	3610-3640
amines	N-H	3300-3500
aromatic rings	C-H	3000-3100
alkenes	C-H	3020-3080
alkanes	C-H	2850-2960
nitriles	C=-N	2210-2260
carbonyl	C=O	1650-1750
amines	C-N	1180-1360

INFRARED ABSORPTION SPECTROMETERS

This document describes dispersive and Fourier-transform spectrometers that are used in infrared absorption spectroscopy.

DISPERSIVE INFRARED SPECTROMETERS

Common light sources are tungsten lamps, Nernst glowers, or glowbars. Dispersive IR spectrometers use a grating monochromator to select wavelengths and are commonly used when a single wavelength is desired to monitor the kinetics of a reaction or as a GC or LC detector.

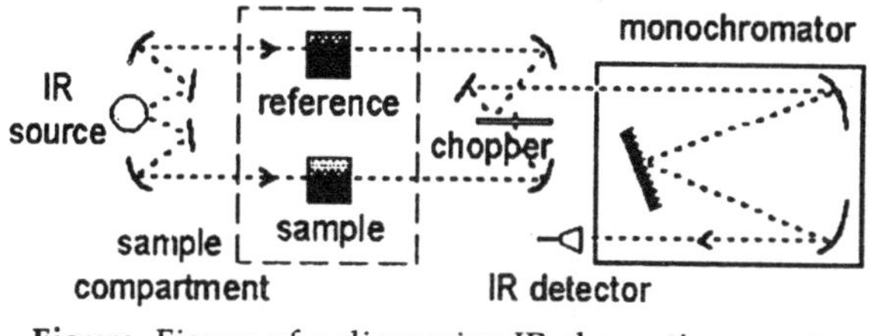

Figure. Figure of a dispersive IR absorption spectrometer

FOURIER-TRANSFORM INFRARED (FTIR)

Modern IR instruments more commonly use Fourier-transform techniques with a Michelson interferometer.

Fig. Figure an FTIR absorption spectrometer

CAVITY-RINGDOWN LASER ABSORPTION SPECTROSCOPY (CRLAS)

Cavity-ringdown laser absorption spectroscopy (CRLAS) is an ultrasensitive method to make quantitative absorption measurements of very low concentrations of analytes. The technique uses a laser pulse that is reflected back and forth between two highly reflecting mirrors. This procedure results in a very long path length, to which the measured absorbance is directly proportional as described by the Beer-Lambert law.

An optical detector is placed behind one of the mirrors to detect the small amount of the light that passes through the mirror. With no absorbing analyte present, the laser pulse will decrease in intensity after each round trip due to the loss of light through the monitoring mirror and other losses. When an absorbing species is present between the mirrors, the intensity of the laser pulse decreases more rapidly. The analyte concentration is determined by calibrating this decay time with known concentration of analyte.

CRLAS can be used from the near-UV to the mid-IR. The wavelength range is only limited by instrumental constraints, such as the availability of high reflectivity mirrors and suitable pulsed laser sources. Accessing the near and mid-IR is accomplished with frequency-conversion techniques such as Raman shifting or optical parametric oscillators.

In principle, the analyte can be in a solid, liquid, or gas. In practice, CRALS is primarily used to measure gas-phase species due to scattering losses in solids or liquids.

INTRACAVITY-ABSORPTION SPECTROSCOPY

Intracavity-absorption spectroscopy is an ultrasensitive method for measuring very small absorptions. Some examples are very low concentrations of analytes, and very weak absorption such as infrared overtone bands. The technique takes advantage of the very high light intensities within a laser resonance cavity, and the sensitivity of the lasing output to losses within the cavity. Small losses within the laser cavity due to absorption results in large changes in the intensity of the laser output.

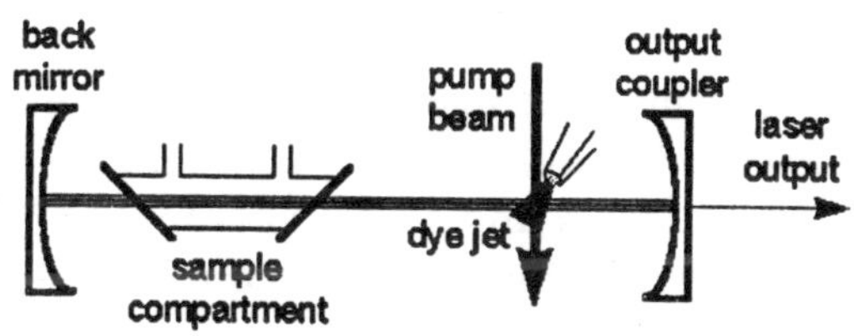

Fig. Schematic of an intracavity-absorption experiment

To record a spectrum the laser output must be tunable, therefore the most common lasers used in intracavity-absorption measurements are dye lasers.

Samples can be solids, liquids, or gases. Liquids and gases are contained in a sample holder that has windows at Brewster's angle to minimize losses due to reflection.

Small absorptions are detected by monitoring the output of the laser beam with a detector. Since the laser light is relatively intense, the main requirement for the detector is that it not be destroyed by the laser beam.

LASER-INDUCED FLUORESCENCE (LIF)

Laser-induced fluorescence (LIF) is the optical emission from molecules that have been excited to higher energy levels by absorption of electromagnetic radiation. The main advantage of fluorescence detection compared to absorption measurements is the greater sensitivity achievable because the fluorescence signal has a very low background. For molecules that can be resonant excitated, LIF provides selective excitation of the analyte to avoid interferences. LIF is useful to study the electronic structure of molecules and to make quantitative measurements of analyte concentrations. Analytical applications include monitoring gas-phase concentrations in the atmosphere, flames, and plasmas; and remote sensing using light detection and ranging (LIDAR). Because of the differences in the nature of the energy-level structure between atoms and molecules, the discussion on atomic fluorescence spectroscopy is in a separate document.

The excitation source for molecular LIF is typically a tunable dye laser in the visible spectral region. Studies in the near-ultraviolet and near-infrared are becoming more common as near-infrared lasers and frequency-doubling methods improve. High-resolution studies require cooling of the molecules to remove spectral congestion and to reduce the Doppler width of the transitions. A separate document on high-resolution spectroscopy describes cooling methods such as molecular beams, free-jet expansions, and cryogenic glass or crystalline matrices.

RAMAN SPECTROSCOPY

Raman spectroscopy is the measurement of the wavelength and intensity of inelastically scattered light from molecules. The Raman scattered light occurs at wavelengths that are shifted from the incident light by the energies of molecular vibrations. The mechanism of Raman scattering is different from that of infrared absorption, and Raman and IR spectra provide complementary information. Typical applications are in structure determination, multicomponent qualitative analysis, and quantitative analysis.

THEORY

The Raman scattering transition moment is:

$$R = < X_i \mid a \mid X_j >$$

where X_i and X_j are the initial and final states, respectively, and a is the polarizability of the molecule:

$$a = a_o + (r-r_e)(da/dr) +\dots \text{ higher terms}$$

where r is the distance between atoms and a_o is the polarizability at the equilibrium bond length, r_e. Polarizability can be defined as the ease of which an electron cloud can be distorted by an external electric field. Since a_o is a constant and $< X_i \mid X_j > = 0$, R simplifies to:

$$R = < X_i \mid (r-r_e)(da/dr) \mid X_j >$$

The result is that there must be a change in polarizability during the vibration for that vibration to inelastically scatter radiation.

Examples of Raman active and inactive vibrations in CO_2

The polarizability depends on how tightly the electrons are bound to the nuclei. In the symmetric stretch the strength of electron binding is different between the minimum and maximum internuclear distances. Therefore the polarizability changes during the vibration and this vibrational mode scatters Raman light (the vibration is Raman active). In the asymmetric stretch the electrons are more easily polarized in the bond that expands but are less easily polarized in the bond that compresses. There is no overall change in polarizability and the asymmetric stretch is Raman inactive.

Raman line intensities are proportional to:

$$nu^4 * \text{sigma}(nu) * I * \exp(-E_i/kT) * C$$

where nu is the frequency of the incident radiation, sigma(nu) is the Raman cross section (typically 10^{-29} cm^2), I is the radiation intensity, $\exp(-E_i/kT)$ is the Boltzmann factor for state i, and C is the analyte concentration.

The most common light source in Raman spectroscopy is an Ar-ion laser. Resonance Raman spectroscopy requires tunable radiation and sources are Ar-ion-laser-pumped cw dye lasers, or high-repetition-rate excimer-laser-pumped pulsed dye lasers. Because Raman scattering is a weak process, a key requirement to obtain Raman spectra is that the spectrometer provide a high rejection of scattered laser light. New methods such as very narrow rejection filters and Fourier-transform techniques are becoming more widespread.

RESONANCE-IONIZATION SPECTROSCOPY (RIS)

Resonance-ionization spectroscopy (RIS) is a spectroscopic method that uses resonant laser excitation to promote an atom or molecule above its ionization potential to create an ion. The ions are detected as a current by a biased collector or by a mass spectrometer. RIS is useful to study the electronic structure of atoms or molecules and to make quantitative measurements of analyte concentrations. For more information see the document on RIMS.

LASER-ENHANCED IONIZATION

Laser-enhanced ionization (LEI) ionizes atoms in a flame. It relies on resonant laser excitation of an atom to a higher electronic excited state. The excited atom is then ionized by collisions in the flame. The ions are collected by biased wires or plates in or near the flame. The LEI method produces lower limits of detection for several elements.

NEAR-INFRARED ABSORPTION SPECTROSCOPY (NIR)

NIR spectroscopy is the measurement of the wavelength and intensity of the absorption of near-infrared light by a sample. Near-infrared light spans the 800 nm - 2.5 μm (12,500 - 4000 cm^{-1}) range and is energetic enough to excite overtones and combinations of molecular vibrations to higher energy levels. NIR spectroscopy is typically used for quantitative measurement of organic functional groups, especially O-H, N-H, and C=O. Detection limits are typically 0.1percent and applications include pharmaceutical, agricultural, polymer, and clinical analysis.

The components and design of NIR instrumentation are similar to uv-vis absorption spectrometers. The light source is usually a tungsten lamp and the detector is usually a PbS solid-state detector. Sample holders can be glass or quartz and typical solvents are CCl_4 and CS_2. The convenient instrumentation of NIR spectroscopy compared to IR spectroscopy makes it much more suitable for on-line monitoring and process control.

POLARIMETRY

Anisotropic crystalline solids, and samples containing an excess of one enantiomer of a chiral molecule, can rotate the orientation of plane-polarized light. Such substances are said to have optical activity. Measurement of this change in polarization orientation is called polarimetry, and the measuring instrument is called a polarimeter. These measurements are useful for studying the structure of anisotropic materials, and for checking the purity of chiral mixtures. A sample that contains only one enantiomer of a chiral molecule is said to be optically pure. The enantiomer that rotates light to the right, or clockwise when viewing in the direction of light propagation, is called the dextrorotatory (d) or (+) enantiomer, and the enantiomer that rotates light to the left, or counterclockwise, is called the levorotatory (l) or (-) enantiomer.

Optical rotation occurs because optically active samples have different refractive indices for left- and right-circularly polarized light. Another way to make this statement is that left- and right-circularly polarized light travel through an optically active sample at different velocities. This condition occurs because a chiral centre has a specific geometric arrangement of four different substituents, each of which has a different electronic polarizability. Light travels through matter by interacting with the electron clouds that are present. Left-circularly polarized light therefore interacts with an anisotropic medium differently than does right-circularly polarized light.

Linearly or plane-polarized light is the superposition of equal intensities of left- and right-circularly polarized light. As plane-polarized light travels through an optically active sample, the left- and right-circularly polarized components travel at different velocities. This difference in velocities creates a phase shift between the two circularly polarized components when they exit the sample. Summing the two components still produces linearly polarized light, but at a different orientation from the light entering the sample.

The simplest polarimeter consists of a monochromatic light source, a polarizer, a sample cell, a second polarizer, which is called the analyser, and a light detector. The analyser is oriented 90° to the polarizer so that no light reaches the detector.

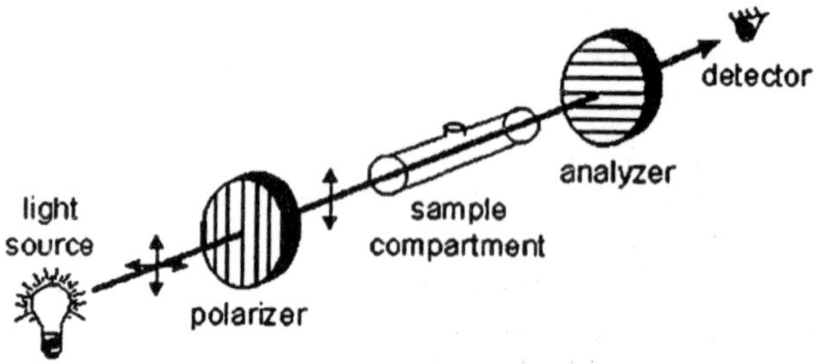

Fig. Schematic of a polarimeter

When an optically active substance is present in the beam, it rotates the polarization of the light reaching the analyser so that there is a component that reaches the detector. The angle that the analyser must be rotated to return to the minimum detector signal is the optical rotation, α.

The amount of optical rotation depends on the number of optically active species through which the light passes, and thus depends on both the sample path length and the analyte concentration. Specific rotation, [α], provides a normalize quantity to correct for this dependence, and is defined as:

$$[\alpha] = \frac{\alpha}{1 + d}$$

where α is the measured optical rotation in degrees, l is the sample path length in decimeters (dm), and d is the density if the sample is pure liquid, or the concentration if the sample is a solution. In either case, the units of d are g/cm^3.

The optical rotatory dispersion (ORD) is the optical rotation as a function of wavelength. It is recorded using a spectropolarimeter, which has a tungsten lamp and a scanning monochromator as the light source. A motorized mount rotates the analyser to maintain a minimum signal at the detector. Usually a modulation is introduced into the polarization angle of the light beam, so that a DC signal to the analyser motor then keeps the detector signal centred at the minimum value.

OPTICAL MATERIALS

This document provides data on materials that are used for optical components (mirrors, lenses, and windows) in different parts of the electromagnetic spectrum.

MIRRORS

X-rays	
Ultraviolet	aluminium
Visible	aluminium
Near infrared	gold
infrared	copper, gold

LENSES

X-rays	
Ultraviolet	fused silica (synthetic quartz), sapphire
Visible	glass, sapphire
Near infrared	glass, sapphire
infrared	CaF_2, ZnSe

WINDOWS

X-rays	beryllium
Ultraviolet	fused silica, sapphire
Visible	glass, sapphire

Near infrared glass, sapphire

infrared NaCl, BaF_2, CaF_2, ZnSe

Refractive index of some optical materials

Medium	**n***
air	1.0003
water	1.333
CaF_2	1.434
BaF_2	1.474
glass	1.5-1.9
synthetic quartz (fused silica)	1.458
crystalline quartz	1.544 (n_o), 1.553 (n_e)
calcite	1.658 (n_o), 1.486 (n_e)
sapphire	1.769
diamond	2.417
silicon	3.478 (1.55 μm)
ZnSe	2.624
	2.403 (10.6 μm)

*Refractive index at 589.3 nm and room temperature, unless noted otherwise.

MASS SPECTROMETRY

INTRODUCTION

Mass spectrometers use the difference in mass-to-charge ratio (m/e) of ionized atoms or molecules to separate them from each other. Mass spectrometry is therefore useful for quantitation of atoms or molecules and also for determining chemical and structural information about molecules. Molecules have distinctive fragmentation patterns that provide structural information to identify structural components.

The general operation of a mass spectrometer is:

1. Create gas-phase ions
2. Separate the ions in space or time based on their mass-to-charge ratio

3. Measure the quantity of ions of each mass-to-charge ratio

The ion separation power of a mass spectrometer is described by the resolution, which is defined as:

$R = m / \Delta m,$

where Δ m is the ion mass and m is the difference in mass between two resolvable peaks in a mass spectrum. E.g., a mass spectrometer with a resolution of 1000 can resolve an ion with a m/e of 100.0 from an ion with an m/e of 100.1.

In general a mass spectrometer consists of an ion source, a mass-selective analyser, and an ion detector. Since mass spectrometers create and manipulate gas-phase ions, they operate in a high-vacuum system. The magnetic-sector, quadrupole, and time-of-flight designs also require extraction and acceleration ion optics to transfer ions from the source region into the mass analyser. The details of mass analyser designs are discussed in the individual documents listed below. Basic descriptions of sample introduction/ionization and ion detection are discussed in separate documents on ionization methods and ion detectors, respectively.

MASS ANALYSER DESIGNS:

- Fourier-transform MS
- Ion-trap MS
- Magnetic-sector MS
- Quadrupole MS
- Time-of-flight MS

MASS SPECTROMETRY IONIZATION METHODS

CHEMICAL IONIZATION (CI)

CI uses a reagent ion to react with the analyte molecules to form ions by either a proton or hydride transfer:

$$MH + C_2H_5^+ \rightarrow MH_2^+ + C_2H_4$$

$$MH + C_2H_5^+ \rightarrow M^+ + C_2H_6$$

The reagent ions are produced by introducing a large excess of methane (relative to the analyte) into an electron impact (EI) ion source. Electron collisions produce CH_4^+ and CH_3^+ which further react with methane to form CH_5^+ and $C_2H_5^+$:

$$CH_4^+ + CH_4 \rightarrow CH_5^+ + CH_3$$

$$CH_3^+ + CH_4 \rightarrow C_2H_5^+ + H_2$$

PLASMA AND GLOW DISCHARGE

A plasma is a hot, partially-ionized gas that effectively excites and ionizes atoms.

A glow discharge is a low-pressure plasma maintained between two electrodes. It is particularly effective at sputtering and ionizing material from solid surfaces.

ELECTRON IMPACT (EI)

An EI source uses an electron beam, usually generated fron a tungsten filament, to ionize gas-phase atoms or molecules. An electron from the beam knocks an electron off of analyte atoms or molecules to create ions.

ELECTROSPRAY IONIZATION (ESI)

The ESI source consists of a very fine needle and a series of skimmers. A sample solution is sprayed into the source chamber to form droplets. The droplets carry charge when the exit the capillary and, as the solvent evapourates, the droplets disappear leaving highly charged analyte molecules. ESI is particularly useful for large biological molecules that are difficult to vapourize or ionize.

FAST-ATOM BOMBARDMENT (FAB)

In FAB a high-energy beam of netural atoms, typically Xe or Ar, strikes a solid sample causing desorption and ionization. It is used for large biological molecules that are difficult to get into the gas phase. FAB causes little fragmentation and usually gives a large molecular ion peak, making it useful for molecular weight determination.

The atomic beam is produced by accelerating ions from an ion source though a charge-exchange cell. The ions pick up an electron in collisions with netural atoms to form a beam of high energy atoms.

FIELD IONIZATION

Molecules can lose an electron when placed in a very high electric field. High fields can be created in an ion source by applying a high voltage between a cathode and an anode called a field emitter. A field emitter consists of a wire covered with microscopic carbon dendrites, which greatly amplify the effective field at the carbon points.

LASER IONIZATION (LIMS)

A laser pulse ablates material from the surface of a sample, and creates a microplasma that ionizes some of the sample constituents. The laser pulse accomplishes both vapourization and ionization of the sample.

MATRIX-ASSISTED LASER DESORPTION IONIZATION (MALDI)

MALDI is a LIMS method of vapourizing and ionizing large biological molecules such as proteins or DNA fragments. The biological molecules are dispersed in a solid matrix such as nicotinic acid.

A UV laser pulse ablates the matrix which carries some of the large molecules into the gas phase in an ionized form so they can be extracted into a mass spectrometer.

PLASMA-DESORPTION IONIZATION (PD)

Decay of ^{252}Cf produces two fission fragments that travel in opposite directions. One fragment strikes the sample knocking out 1-10 analyte ions. The other fragment strikes a detector and triggers the start of data acquisition. This ionization method is especially useful for large biological molecules.

RESONANCE IONIZATION (RIMS)

One or more laser beams are tuned in resonance to transistions of a gas-phase atom or molecule to promote it in a stepwise fashion above its ionization potential to create an ion. Solid samples must be vapourized by heating, sputtering, or laser ablation.

SECONDARY IONIZATION (SIMS)

A primary ion beam; such as $^{3}He^{+}$,$^{16}O^{+}$, or $^{40}Ar^{+}$; is accelerated and focused onto the surface of a sample and sputters material into the gas phase. Approximately 1percent of the sputtered material comes off as ions, which can then be analysed by a mass spectrometer. SIMS has the advantage that material can be continually sputtered from a surface to determine analyte concentrations as a function of distance from the original surface (depth profiling).

SPARK SOURCE

A spark source ionizes analytes in solid samples by pulsing an electric current across two electrodes. If the sample is a metal it can serve as one of the electrodes, otherwise it can be mixed with graphite and placed in a cup-shaped electrode.

THERMAL IONIZATION (TIMS)

Thermal ionization is used for elemental or refractory materials. A sample is deposited on a metal ribbon, such as Pt or Re, and an electric current heats the metal to a high temperature. The ribbon is often coated with graphite to provide a reducing effect.

ION DETECTORS

CHANNELTRON

A channeltron is a horn-shaped continuous dynode structure that is coated on the inside with a electron emissive material. An ion striking the channeltron creates secondary electrons that have an avalanche effect to create more secondary electrons and finally a current pulse.

DALY DETECTOR

A Daly detector consists of a metal knob that emits secondary electrons when struck by an ion. The secondary electrons are accelerated onto a scintillator that produces light that is then detected by a photomultiplier tube.

ELECTRON MULTIPLIER TUBE (EMT)

Electron multiplier tubes are similar in design to photomultiplier tubes. They consist of a series of biased dynodes that eject secondary electrons when they are struck by an ion. They therefore multiply the ion current and can be used in analog or digital mode.

FARADAY CUP

A Faraday cup is a metal cup that is placed in the path of the ion beam. It is attached to an electrometer, which measures the ion-beam current. Since a Faraday cup can only be used in an analog mode it is less sensitive than other detectors that are capable of operating in pulse-counting mode.

MICROCHANNEL PLATE

A microchannel plate consists of an array of glass capillaries (10-25 um inner diameter) that are coated on the inside with a electron-emissive material. The capillaries are biased at a high voltage and like the channeltron, an ion that strikes the inside wall one of the capillaries creates an avalanche of secondary electrons. This cascading effect creates a gain of 10^3 to 10^4 and produces a current pulse at the output.

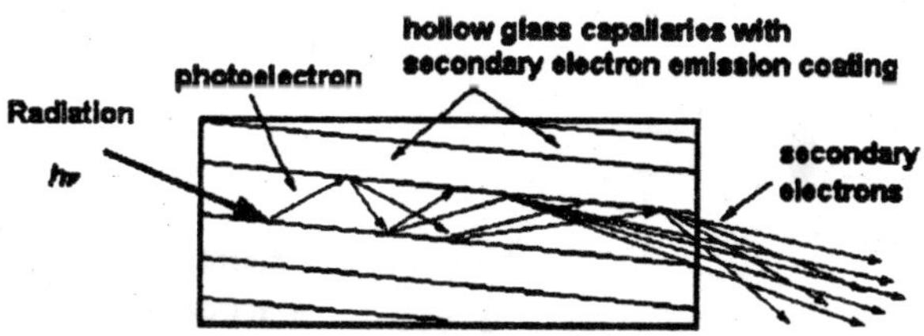

Fig. Schematic of a microchannel plate

Microchannel plates (MCP) are also used as an intensifier for low-intensity light detection with array detectors.

FOURIER-TRANSFORM MASS SPECTROMETRY

Fourier-transform mass spectrometry takes advantage of ion-cyclotron resonance to select and detect ions.

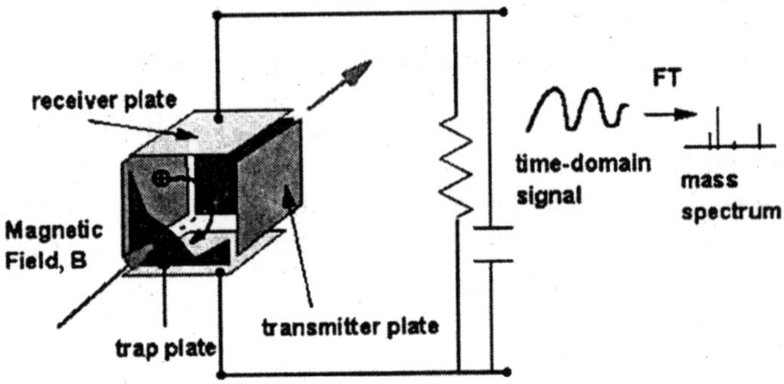

Fig. Schematic of a FT-MS

ION-TRAP MASS SPECTROMETRY

The ion-trap mass spectrometer uses three electrodes to trap ions in a small volume. The mass analyser consists of a ring electrode separating two hemispherical electrodes. A mass spectrum is obtained by changing the electrode voltages to eject the ions from the trap. The advantages of the ion-trap mass spectrometer include compact size, and the ability to trap and accumulate ions to increase the signal-to-noise ratio of a measurement.

MAGNETIC-SECTOR MASS SPECTROMETRY

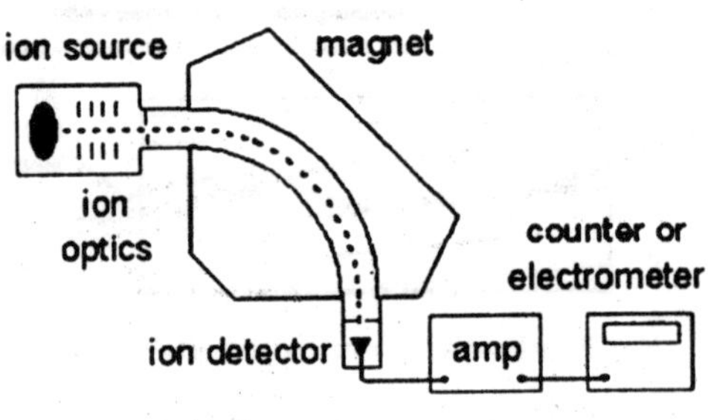

Fig. Schematic of a magnetic-sector mass spectrometer

THEORY

The ion optics in the ion-source chamber of a mass spectrometer extract and accelerate ions to a kinetic energy given by:

K.E. = $0.5\ mv^2 = eV$

where m is the mass of the ion, v is it's velocity, e is the charge of the ion and V is the applied voltage of the ion optics.

The ions enter the flight tube between the poles of a magnet and are deflected by the magnetic field, H. Only ions of mass-to-charge ratio that have equal centrifugal and centripetal forces pass through the flight tube:

$mv^2 / r = Hev$

centrifugal = centripetal forces.

Where r is the radius of curvature of the ion path:

$r = mv / eH$

$$= \frac{1}{H}\sqrt{2V\frac{m}{e}}$$

$$\frac{m}{e} = \frac{H^2 r^2}{2V}$$

This equation shows that the m/e of the ions that reach the detector can be varied by changing either H or V.

Single Focusing analysers:

A circular beam path of 180, 90, or 60 degrees can be used. The various forces influencing the particle separate ions with different mass-to-charge ratios.

Double Focusing analysers:

An electrostatic analyser is added in this type of instrument to separate particles with difference in kinetic energies.

QUADRUPOLE MASS SPECTROMETRY

A quadrupole mass filter consists of four parallel metal rods arranged as in the figure. Two opposite rods have an applied potential of (U+Vcos(wt)) and the other two rods have a potential of -(U+Vcos(wt)), where U is a dc voltage and Vcos(wt) is an ac voltage. The applied voltages affect the trajectory of ions traveling down the flight path centred between the four rods. For given dc and ac voltages, only ions of a certain mass-to-charge ratio pass through the quadrupole filter and all other ions are thrown out of their original path. A mass spectrum is obtained by monitoring the ions passing through the quadrupole filter as the voltages on the rods are varied. There are two methods: varying w and holding U and V constant, or varying U and V (U/V) fixed for a constant w.

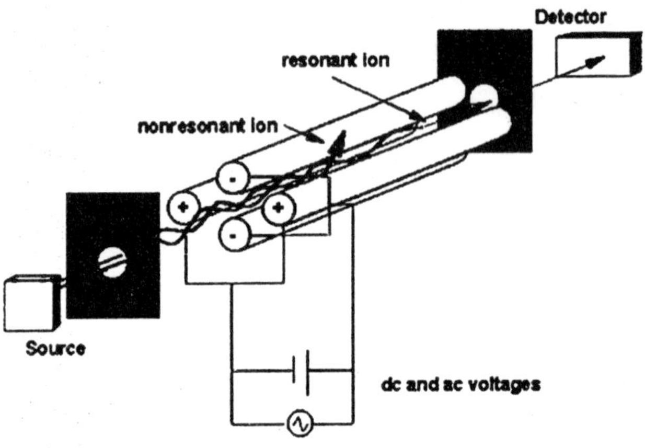

Fig. Schematic of a quadrupole filter

Quadrupole mass spectrometers consist of an ion source, ion optics to accelerate and focus the ions through an aperture into the quadrupole filter, the quadrupole filter itself with control voltage supplies, an exit aperture, an ion detector and electronics, and a high-vacuum system.

TIME-OF-FLIGHT MASS SPECTROMETRY (TOF-MS)

A time-of-flight mass spectrometer uses the differences in transit time through a drift region to separate ions of different masses. It operates in a pulsed mode so ions must be produced or extracted in pulses. An electric field accelerates all ions into a field-free drift region with a kinetic energy of qV, where q is the ion charge and V is the applied voltage. Since the ion kinetic energy is $0.5mv^2$, lighter ions have a higher velocity than heavier ions and reach the detector at the end of the drift region sooner.

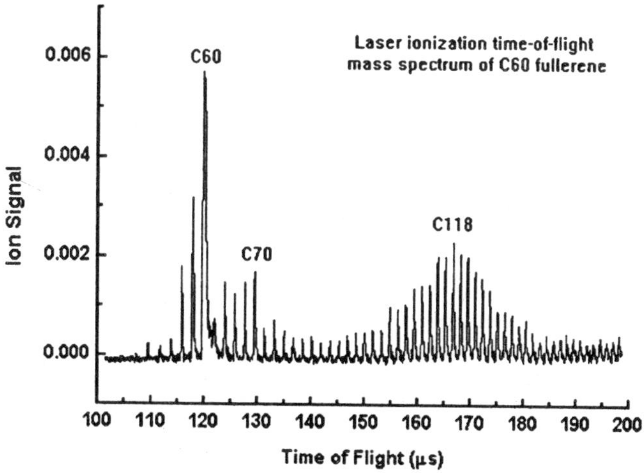

Fig. Example of a TOF mass spectrum

The mass of C60 is 720 amu (fullerene mass spectrum courtesy of Craig Watson, Virginia Tech).

THEORY

$$K.E. = qV$$

$$\frac{1}{2} mv^2 = qV$$

$$V = \left(\frac{2qV}{m}\right)^{1/2}$$

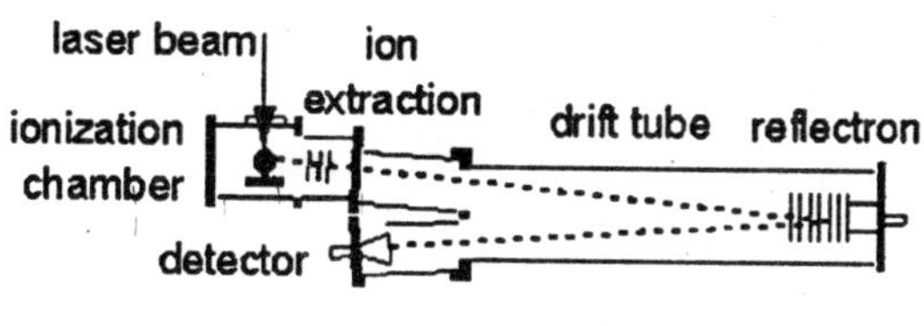

Fig. Schematic of a reflectron TOF-MS

The transit time (t) through the drift tube is L/V where L is the length of the drift tube.

$t = L / (2V/_{m/q})^{1/2}$

This schematic shows ablation of ions from a solid sample with a pulsed laser. The reflectron is a series of rings or grids that act as an ion mirror. This mirror compensates for the spread in kinetic energies of the ions as they enter the drift region and improves the resolution of the instrument. The output of an ion detector is displayed on an oscilloscope as a function of time to produce the mass spectrum.

ULTRAVIOLET AND VISIBLE ABSORPTION SPECTROSCOPY (UV-VIS)

UV-vis spectroscopy is the measurement of the wavelength and intensity of absorption of near-ultraviolet and visible light by a sample. Ultraviolet and visible light are energetic enough to promote outer electrons to higher energy levels. UV-vis spectroscopy is usually applied to molecules and inorganic ions or complexes in solution. The uv-vis spectra have broad features that are of limited use for sample identification but are very useful for quantitative measurements. The concentration of an analyte in solution can be determined by measuring the absorbance at some wavelength and applying the Beer-Lambert Law.

The light source is usually a hydrogen or deuterium lamp for uv measurements and a tungsten lamp for visible measurements. The wavelengths of these continuous light sources are selected with a wavelength separator such as a prism or grating monochromator. Spectra are obtained by scanning the wavelength separator and quantitative measurements can be made from a spectrum or at a single wavelength.

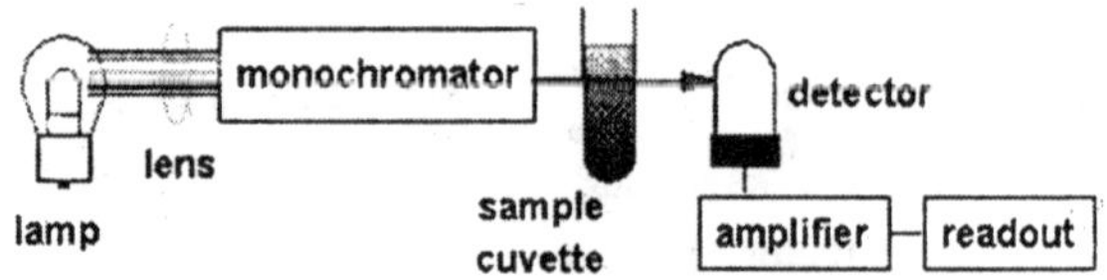

Fig. Schematic of a single beam uv-vis spectrophotometer

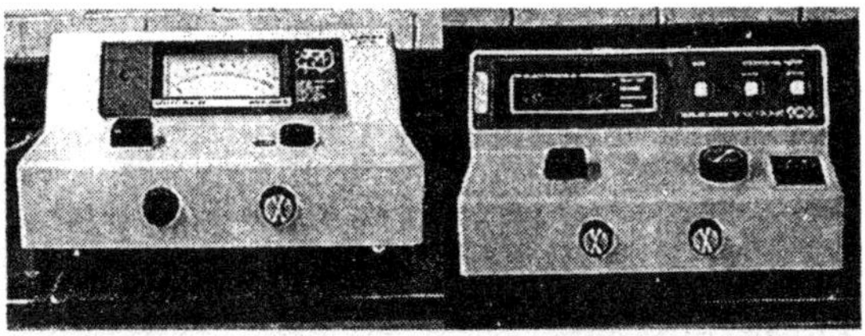

Fig. Pictures of single beam uv-vis spectrophotometers (Spectronic 20 and 20D)

Larger picture of the Spectronic 20D with labels.

DUAL-BEAM UV-VIS SPECTROPHOTOMETER

In single-beam uv-vis absorption spectroscopy, obtaining a spectrum requires manually measuring the transmittance of the sample and solvent at each wavelength. The double-beam design greatly simplifies this process by measuring the transmittance of the sample and solvent simultaneously. The detection electronics can then manipulate the measurements to give the absorbance.

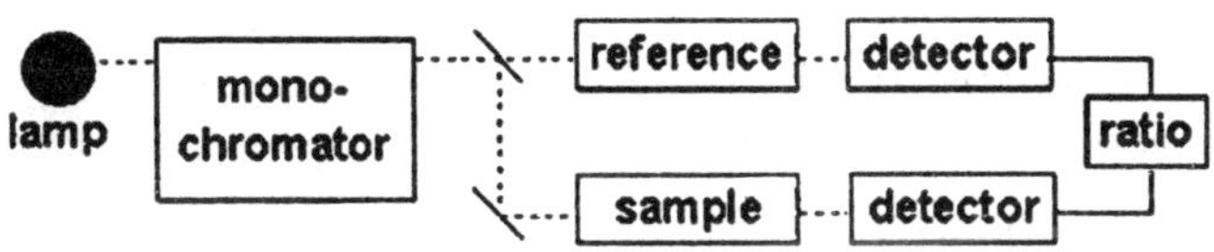

Fig. Schematic of a dual-beam uv-vis spectrophotometer

ELECTRON PARAMAGNETIC RESONANCE (EPR, ESR) SPECTROSCOPY

When an atom or molecule with an unpaired electron is placed in a magnetic field, the spin of the unpaired electron can align either in the same direction or in the opposite direction as the field. These two electron alignments have different energies and application of a magnetic field to an unpaired electron lifts the degeneracy of the ±1/2 spins of the electron.

Electron-paramagnetic-resonance (EPR) or electron-spin-resonance (ESR) spectroscopy measures the absorption of microwave radiation by an unpaired electron when it is placed in a strong magnetic field.

Species that contain unpaired electrons:

1. Free radicals
2. Odd electron molecules
3. Transition-metal complexes
4. Lanthanide ions
5. Triplet-state molecules

A klystron tube generates monochromatic microwave radiation (~9500 MHz) in an EPR instrument. The microwave radiation travels down a waveguide to the sample which is held between magnets.

Spectra are obtained by measuring the absorption of the microwave radiation while scanning the magnetic-field strength. EPR spectra are usually displayed in derivative form to improve the signal-to-noise ratio.

NUCLEAR MAGNETIC RESONANCE (NMR) SPECTROSCOPY

Nuclei with an odd number of protons, neutrons, or both, will have an instrinsic nuclear spin.

Spin quantum number for various nuclei

Number of protons	Number of Neutrons	Spin Quantum Number	Examples
Even	Even	0	^{12}C, ^{16}O, ^{32}S
Odd	Even	1/2	^{1}H, ^{19}F, ^{31}P
"	"	3/2	^{11}B,^{35}Cl, ^{79}Br
Even	Odd	1/2	^{13}C
"	"	3/2	^{127}I
"	"	5/2	^{17}O
Odd	Odd	1	^{2}H, ^{14}N

When a nucleus with a non-zero spin is placed in a magnetic field, the nuclear spin can align in either the same direction or in the opposite direction as the field. These two nuclear spin alignments have different energies and application of a magnetic field lifts the degeneracy of the nuclear spins. A nucleus that has its spin aligned with the field will have a lower energy than when it has its spin aligned in the opposite direction to the field.

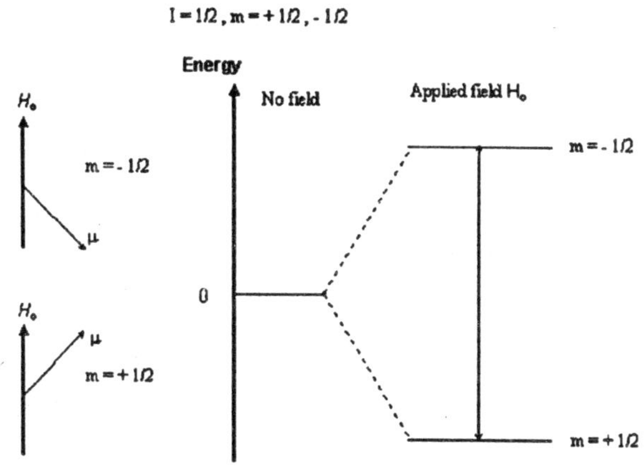

Nuclear magnetic resonance (NMR) spectroscopy is the absorption of radiofrequency radiation by a nucleus in a strong magnetic field. Absorption of the radiation causes the nuclear spin to realign or flip in the higher-energy direction. After absorbing energy the nuclei will reemit RF radiation and return to the lower-energy state.

The energy of a NMR transition depends on the magnetic-field strength and a proportionality factor for each nucleus called the magnetogyric ratio. The local environment around a given nucleus in a molecule will slightly perturb the local magnetic field exerted on that nucleus and affect its exact transition energy. This dependence of the transition energy on the position of a particular atom in a molecule makes NMR spectroscopy extremely useful for determining the structure of molecules.

There are two NMR spectrometer designs, continuous-wave (cw), and pulsed or Fourier-transform (FT-NMR). CW-NMR spectrometers have largely been replaced with pulsed FT-NMR instruments. However due to the lower maintenance and operating cost of cw instruments, they are still commonly used for routine ^{1}H NMR spectroscopy at 60 MHz. (Low-resolution cw instruments require only water-cooled electromagnets instead of the liquid-He-cooled superconducting magnets found in higher-field FT-NMR spectrometers.) These two spectrometer designs are described in separate CW-NMR and FT-NMR documents.

CONTINUOUS-WAVE NUCLEAR MAGNETIC RESONANCE (NMR) SPECTROSCOPY

Continuous-wave NMR spectrometers have largely been replaced with pulsed FT-NMR instruments. However due to the lower maintenance and operating cost of cw instruments, they are still commonly used for routine ^{1}H NMR spectroscopy at 60 MHz. (Low-resolution cw instruments require only water-cooled electromagnets instead of the liquid-He-cooled superconducting magnets found in higher-field FT-NMR spectrometers.)

A continuous wave -NMR spectrometer consists of a control console, magnet, and two orthogonal coils of wire that serve as antennas for radiofrequency (RF) radiation. One coil is attached to an RF generator and serves as a transmitter. The other coil is the RF pick-up coil and is attached to the detection electronics.

Since the two coils are orthogonal, the pick-up coil cannot directly receive any radiation from the generator coil. When a nucleus absorbs RF radiation, it can become reoriented due to its normal movement in solution and re-emit the RF radiation is a direction that can be received by the pick-up coil. This orthogonal coil arrangement greatly increases the sensitivity of NMR spectroscopy, similar to optical fluorescence.

Spectra are obtained by scanning the magnet and recording the pick-up coil signal on paper at the control console.

FOURIER-TRANSFORM NUCLEAR MAGNETIC RESONANCE (FT-NMR) SPECTROSCOPY

Fourier-transform NMR spectrometers use a pulse of radiofrequency (RF) radiation to cause nuclei in a magnetic field to flip into the higher-energy alignment. Due to the Heisenberg uncertainty principle, the frequency width of the RF pulse (typically 1-10 μs) is wide enough to simultaneously excite nuclei in all local environments. All of the nuclei will re-emit RF radiation at their respective resonance frequencies, creating an interference pattern in the resulting RF emission versus time, known as a free-induction decay (FID). The frequencies are extracted from the FID by a Fourier transform of the time-based data.

An FT-NMR spectrometer consists of a control console, magnet, and a coil of wire that serves as the antenna for transmitting and receiving the RF radiation. (Only one coil is necessary because signal reception does not begin until after the end of the excitation pulse.) Because the FID results from the emission due to nuclei in all environments, each pulse contains an interference pattern from which the complete spectrum can be obtained. Because of this multiplex (or Fellgett) advantage, repetitive signals can be summed and averaged to greatly improve the signal-to-noise ratio of the resulting FID.

MOSSBAUER SPECTROSCOPY

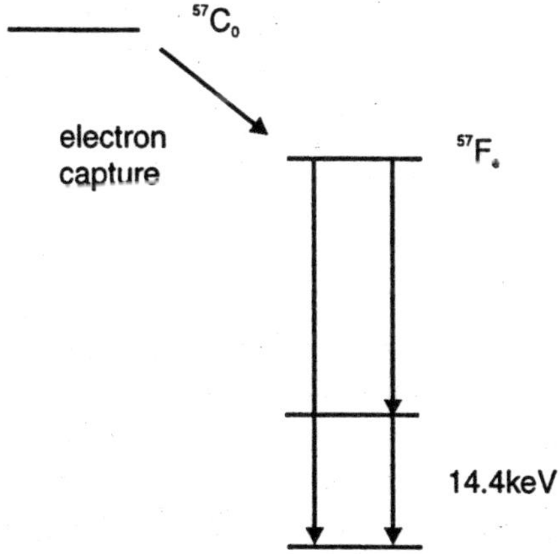

Fig. Production of gamma rays for ^{57}Fe Mossbauer spectroscopy

The Mossbauer effect is the recoil-free emission of gamma radiation from a solid radioactive material. Since the gamma emission is recoil-free, it can be resonantly absorbed by stationary atoms, i.e., also in a solid. The nuclear transitions are very sensitive to the local environment of the atom and Mossbauer spectroscopy is a sensitive probe of the different environments an atom occupies in a solid material.

Approximately 90percent of the ^{57}Fe nuclear excited state decays through the intermediate level to produce 14.4 keV gamma radiation. These gamma photons can then be absorbed by ^{57}Fe in a sample.

The gamma ray source is a radioactive element that is mechanically vibrated back and forth to Doppler shift the energy of the emitted gamma radiation. The schematic below shows a transmission Mossbauer experiment. As the energy of the gamma radiation is scanned by Doppler shifting, the detector records the frequencies of gamma radiation that are absorbed by the sample.

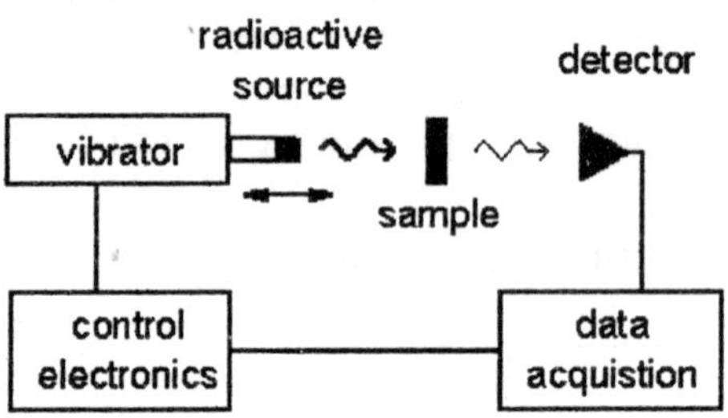

Fig. Schematic of an experimental set-up for transmission Mossbauer spectroscopy

X-RAY FLUORESCENCE

X-ray fluorescence is a spectroscopic method that is commonly used for solids in which secondary x-ray emission is generated by excitation of a sample with x-rays. The x-rays eject inner-shell electrons. Outer-shell electrons take their place and emit photons in the process. The wavelength of the photons depends on the energy difference between the outer-shell and inner-shell electron orbitals. The amount of x-ray fluorescence is very sample dependent and quatitative analysis requires calibration with standards that are similar to the sample matrix.

Solid samples are usually powdered and pressed into a wafer or fused in a borate glass. The sample is then placed in the sample chamber of an XRF spectrometer, and irradiated with a primary X-ray beam. The X-ray fluorescence is recorded with either an X-ray detector after wavelength dispersion or with an energy-dispersive detector.

RADIATION SOURCES FOR ABSORPTION AND FLUORESCENCE SPECTROSCOPY

LAMPS

Lamps convert electrical energy into radiation. Different designs and materials are needed to produce light in different parts of the electromagnetic spectrum. The following sections describe several different types of lamps that are useful in spectroscopy.

BLACKBODY SOURCES

A hot material, such as an electrically-heated filament in a light bulb, emits a continuum spectrum of light. The spectrum is approximated by Planck's radiation law for blackbody radiators:

$$B = \frac{8\pi h v^3}{c^3\left(e^{-hv/kT} - 1\right)}$$

The most common incandescent lamps and their wavelength ranges are:

tungsten filament lamps: 350 nm – 2.5 mm

glowbar: 1 – 40 mm

Nernst glower: 400 nm – 20 mm

Tungsten lamps are used in visible and near-infrared (NIR) absorption spectroscopy and the glowbar and Nernst glower are used for infrared spectroscopy.

DISCHARGE LAMPS

Discharge lamps, such as neon signs, pass an electric current through a rare gas or metal vapour to produce light.

The electrons collide with gas atoms, exciting them to higher energy levels which then decay to lower levels by emitting light. Low-pressure lamps have sharp line emission characteristic of the atoms in the lamp, and high-pressure lamps have broadened lines superimposed on a continuum.

Common discharge lamps and their wavelength ranges are:

hydrogen or deuterium: 160 - 360 nm

mercury: 253.7 nm, and weaker lines in the near-uv and visible

Ne, Ar, Kr, Xe discharge lamps: many sharp lines throughout the near-uv to near-IR

xenon arc: 300 - 1300 nm

Deuterium lamps are the uv source in uv-vis absorption spectrophotometers. The sharp lines of the mercury and rare gas discharge lamps are useful for wavelength calibration of optical instrumentation. Mercury and xenon arc lamps are used to excite fluorescence.

HOLLOW-CATHODE LAMPS

Hollow-cathode lamps are a type of discharge lamp that produce narrow emission from atomic species. They get their name from the cup-shaped cathode, which is made from the element of interest. The electric discharge ionizes rare gas atoms, which are accelerated into the cathode and sputter metal atoms into the gas phase. Collisions with gas atoms or electrons excite the metal atoms to higher energy levels, which decay to lower levels by emitting light.

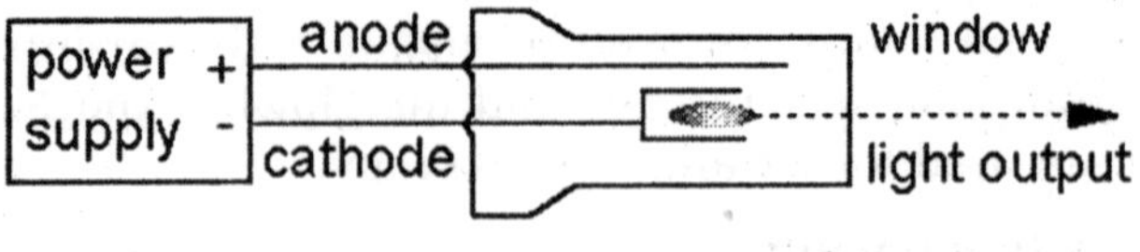

Fig. Schematic of a hollow-cathode lamp

Hollow-cathode lamps have become the most common light source for atomic absorption (AA) spectroscopy. They are also sometimes used as an excitation source for atomic-fluorescence spectroscopy (AFS).

LASERS

A laser is a coherent and highly directional radiation source. LASER stands for Light Amplification by Stimulated Emission of Radiation.

A laser consists of at least three components:

1. A gain medium that can amplify light that passes through it
2. An energy pump source to create a population inversion in the gain medium
3. Two mirrors that form a resonator cavity

The gain medium can be solid, liquid, or gas and the pump source can be an electrical discharge, a flashlamp, or another laser. The specific components of a laser vary depending on the gain medium and whether the laser is operated continuously (cw) or pulsed. The following headings describe specific laser designs.

GAS LASERS

Gas lasers are typically excited by an electrical discharge.

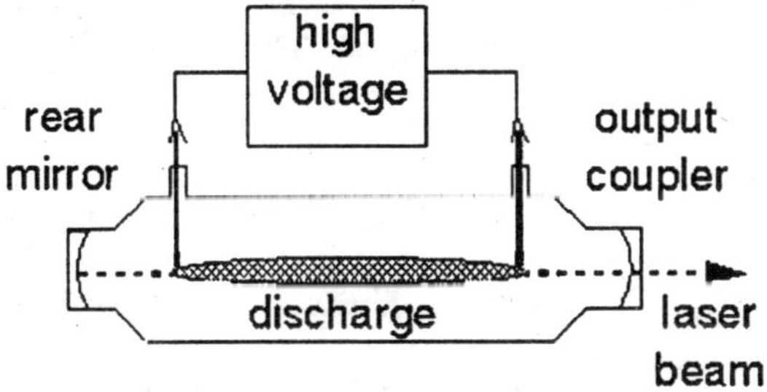

Fig. Schematic of a cw gas laser

Some gas lasers and their dominant lasing wavelength(s):

- Excimer: ArF^* - 248 nm, $XeCl^*$ - 308 nm (pulsed)
- Nitrogen: 337 nm (pulsed)
- He-Ne: 632.8 nm (cw)

- Ar ion: 488, 541 nm (cw)
- CO2: 10.6 μm (cw or pulsed)

SOLID-STATE LASERS

The gain medium in a solid-state laser is an impurity centre in a crystal or glass. Solid-state lasers made from semiconductors are described below. The first laser was a ruby crystal (Cr^{3+} in Al_2O_3) that lased at 694 nm when pumped by a flashlamp. The most commonly used solid-state laser is one with Nd^{3+} in a $Y_3Al_5O_8$ (YAG) or $YLiF_4$(YLF) crystal or in a glass. These Nd^{3+} lasers operate either pulsed or cw and lase at approximately 1064 nm. The high energies of pulsed Nd^{3+}:YAG lasers allow efficient frequency doubling (532 nm), tripling (355 nm), or quadrupling (266 nm), and the 532 nm and 355 nm beams are commonly used to pump tunable dye lasers.

DYE LASERS

The gain medium in a dye laser is an organic dye molecule that is dissolved in a solvent. The dye and solvent are circulated through a cell or a jet, and the dye molecules are excited by flashlamps or other lasers. Pulsed dye lasers use a cell and cw dye lasers typically use a jet. The organic dye molecules have broad fluorescence bands and dye lasers are typically tunable over 30 to 80 nm. Dyes exist to cover the near-uv to near-infrared spectral region: 330 - 1020 nm.

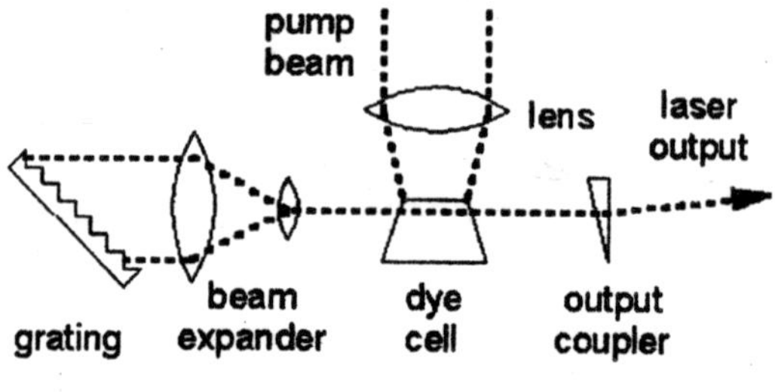

Fig. Schematic of a pulsed dye laser

SEMICONDUCTOR LASERS

Semiconductor lasers are light-emitting diodes within a resonator cavity that is formed either on the surfaces of the diode or externally. An electric current passing through the diode produces light emission when electrons and holes recombine at the p-n junction. Because of the small size of the active medium, the laser output is very divergent and requires special optics to produce a good beam shape. These lasers are used in optical-fiber communications, CD players, and in high-resolution molecular spectroscopy in the near-infrared. Diode laser arrays can replace flashlamps to efficiently pump solid-state lasers. Diode lasers are tunable over a narrow range and different semiconductor materials are used to make lasers at 680, 800, 1300, and 1500 nm.

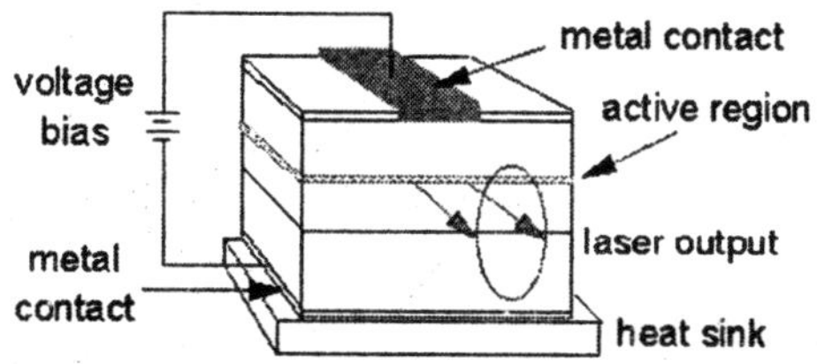

Fig. Schematic of a semiconductor diode laser

FILTERS

Filters separate different parts of the electromagnetic spectrum by absorbing or reflecting certain wavelengths and transmitting other wavelengths.

COLOUR FILTERS

Colour filters are glass substrates containing absorbing species that absorb certain wavelengths. A typical example is a cut-on colour filter, which blocks short wavelength light such as an excitation source, and transmits longer wavelength light such as fluorescence that reaches a detector.

INTERFERENCE FILTERS

Interference filters are made of multiple dielectric thin films on a substrate. They use interference to selectively transmit or reflect a certain range of wavelengths. A typical example is a bandpass interference filter that transmits a narrow range of wavelengths, and can isolate a single emission line from a discharge lamp.

INTERFEROMETERS

The purpose of an interferometer is similar to that of a filter or monochromator, i.e., to isolate a specific portion of the electromagnetic spectrum. Unlike prism or grating monochromators, interferometers are not dispersive instruments, but use interference to selectively transmit a certain wavelength. The links below lead to descriptions of three interferometer designs.

TYPES OF INTERFEROMETERS

Fabry-Perot Interferometer (etalon)

Used in high-resolution applications, such as atomic spectroscopy or measurement of narrow-band laser linewidths.

Michelson Interferometer

Used in fourier-transform infrared absorption spectrometers (FTIR).

Mach-Zender interferometer

Used to measure refractive index changes in gases and in interference microscopes to image transparent samples.

FABRY-PEROT INTERFEROMETER

The Fabry-Perot interferometer is commonly used as a narrow-bandpass filter or as an instrument to measure spectral linewidths.

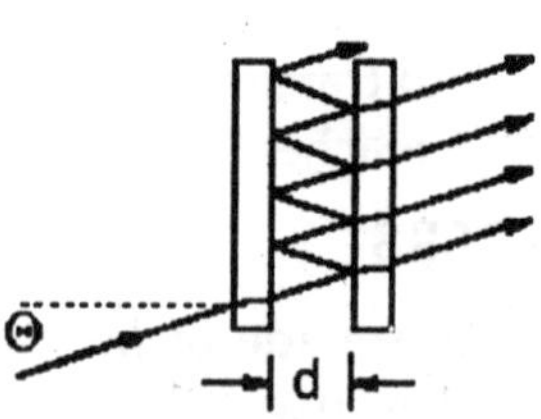

Fig. Schematic of a Fabry-Perot etalon

Fabry-Perot interferometers that cannot be scanned are called etalons.

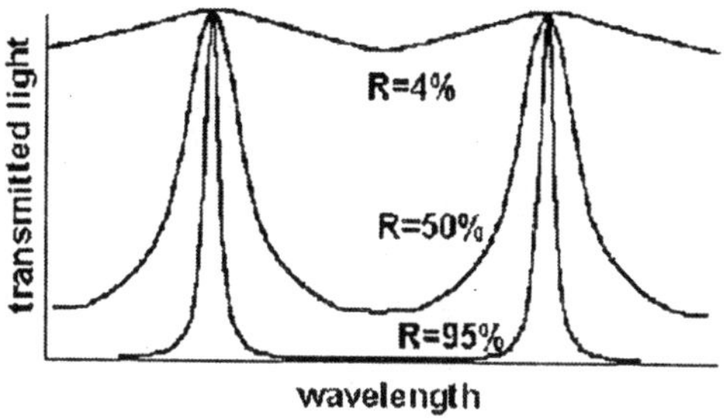

Fig. Transmission through a Fabry-Perot interferometer as a function of wavelength

MICHELSON INTERFEROMETER

The Michelson interferometer design is used in Fourier-transform infrared absorption spectrometers (FTIR).

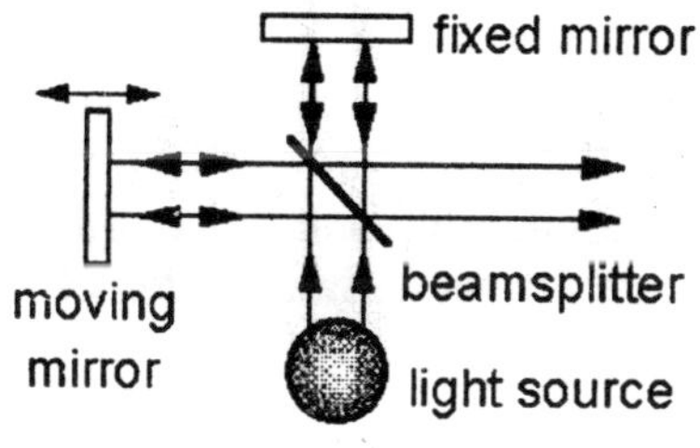

Fig. Schematic of a Michelson interferometer

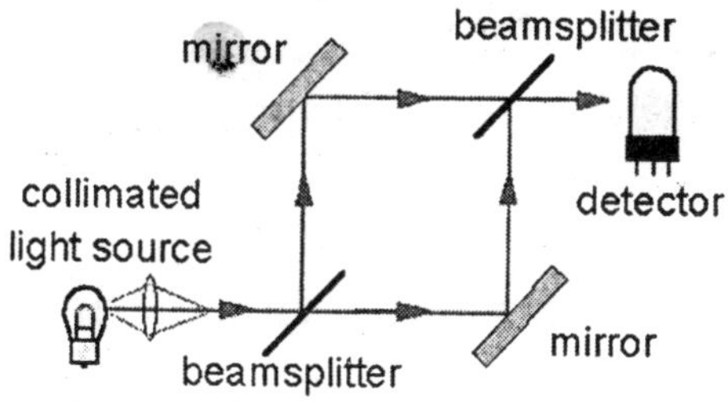

Fig. Schematic of a Mach-Zender interferometer

MACH-ZENDER INTERFEROMETER

The Mach-Zender interferometer design is used to measure refractive index changes in gases, and in interference microscopes to image transparent samples.

OPTICAL RADIATION DETECTORS

Detectors convert light energy to an electrical signal. In spectroscopy, they are typically placed after a wavelength separator to detect a selected wavelength of light. Different types of detectors are sensitive in different parts of the electromagnetic spectrum.

Index of Radiation Detectors

- Charge-coupled detector array (CCD)
- Semiconductors (photodiodes and photovoltaics)
- Photodiode array (PDA)
- Photomultiplier tube (PMT)

CHARGE-COUPLED DEVICES (CCD)

A CCD is an integrated-circuit chip that contains an array of capacitors that store charge when light creates e-hole pairs. The charge accumulates and is read in a fixed time interval. CCDs are used in similar applications to other array detectors such as photodiode arrays, although the CCD is much more sensitive for measurement of low light levels.

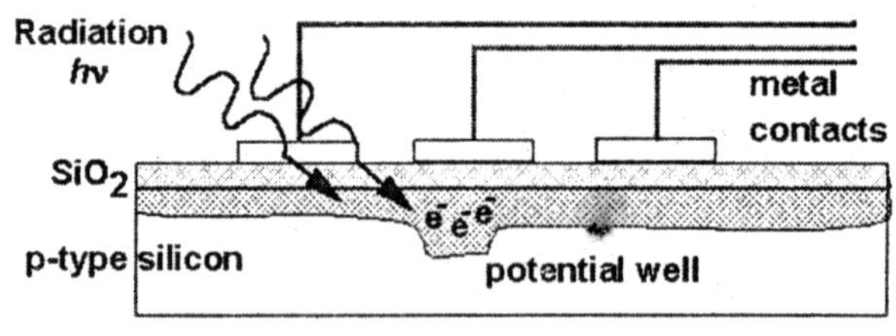

Fig. Schematic of a CCD

PHOTODIODE AND PHOTOVOLTAIC DETECTORS

When a photon strikes a semiconductor, it can promote an electron from the valence band (filled orbitals) to the conduction band (unfilled orbitals) creating an electron(-) - hole(+) pair.

The concentration of these electron-hole pairs is dependent on the amount of light striking the semiconductor, making the semiconductor suitable as an optical detector. There are two ways to monitor the concentration of electron-hole pairs.

In photodiodes, a voltage bias is present and the concentration of light-induced electron-hole pairs determines the current through semiconductor. Photovoltaic detectors contain a p-n junction, that causes the electron-hole pairs to separate to produce a voltage that can be measured.

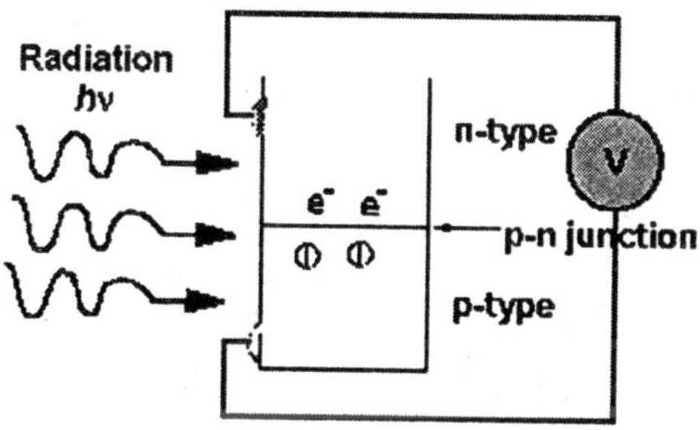

Fig. Schematic of semiconductor detector

Photodiode detectors are not as sensitive as PMTs but they are small and robust.

WAVELENGTH RANGE

Wavelength range

Detector type	λ (μ)
Si	0.2 - 1.1
Ge	0.4 - 1.8
InAs	1.0 - 3.8
InSb	1.0 - 7.0
InSb (77K)	1.0 - 5.6
HgCdTe (77K)	1.0 -25.0

PHOTODIODE ARRAY DETECTORS (PDA)

A photodiode array (PDA) is a linear array of discrete photodiodes on an integrated circuit (IC) chip. For spectroscopy it is placed at the image plane of a spectrometer to allow a range of wavelengths to be detected simultaneously. In this regard it can be thought of as an electronic version of photographic film. Array detectors are especially useful for recording the full uv-vis absorption spectra of samples that are rapidly passing through a sample flow cell, such as in an HPLC detector.

PDAs work on the same principle as simple photovoltaic detectors.

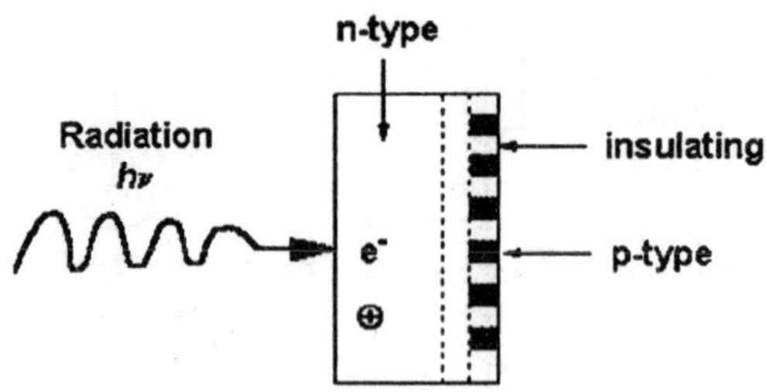

Fig. Schematic of a PDA

Light creates electron-hole pairs and the electrons migrate to the nearest PIN junction. After a fixed integration time the charge at each element is sequentially read with solid-state circuitry to generate the detector response as a function of linear distance along the array. PDAs are available with 512, 1024, or 2048 elements with typical dimensions of ~ 25 μm wide and 1-2 mm high.

PHOTOMULTIPLIER TUBE (PMT)

Photomultiplier Tubes (PMTS) are light detectors that are useful in low intensity applications such as fluorescence spectroscopy. Due to high internal gain, PMTs are very sensitive detectors.

PMTs are similar to phototubes. They consist of a photocathode and a series of dynodes in an evacuated glass enclosure. Photons that strikes the photoemissive cathode emits electrons due to the photoelectric effect. Instead of collecting these few electrons (there should not be a lot, since the primarily use for PMT is for verly low signal) at an anode like in the phototubes, the electrons are accelerated towards a series of additional electrodes called dynodes. These electrodes are each maintained at a more positive potential. Additional electrons are generated at each dynode. This cascading effect creates 10^5 to 10^7 electrons for each photon hitting the first cathode depending on the number of dynodes and the accelerating voltage. This amplified signal is finally collected at the anode where it can be measured.

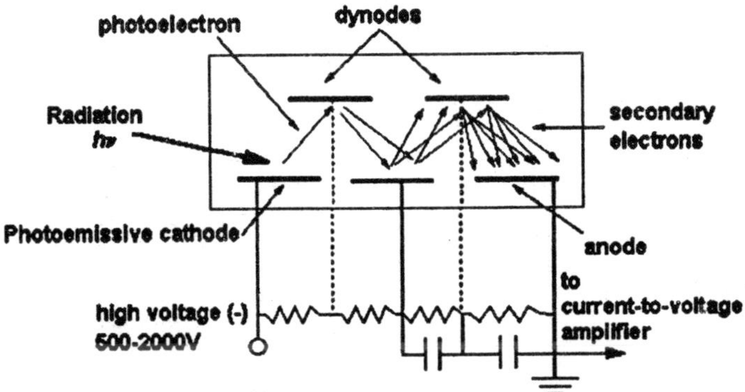

Fig. Schematic of a PMT

TYPICAL SPECIFICATIONS

Wavelength range: 110-1100 nm

(wavelength sensitivity dependent on wavelength, uv-sensitive PMTs must have uv-transmitting windows, see optical materials)

DIFFRACTION

Diffraction is a wave property of electromagnetic radiation that causes the radiation to bend as it passes by an edge or through an aperture. Diffraction effects increase as the physical dimension of the aperture approaches the wavelength of the radiation. Diffraction of radiation results in interference that produces dark and bright rings, lines, or spots, depending on the geometry of the object causing the diffraction. Common interference effects for visible light are the rainbow pattern produced by an oil film on wet pavement and the diffraction of light from a narrow slit or a diffraction grating.

DIFFRACTION METHODS

A certain wavelength of radiation will constructively interfere when partially reflected between surfaces that produce a path difference equal to an integral number of wavelengths. This condition is described by the Bragg law:

$$n\lambda = 2d\sin\theta$$

where n is an integer, lambda is the wavelength of the radiation, d is the spacing between surfaces, and theta is the angle between the radiation and the surfaces. This relation demonstrates that interference effects are observable only when radiation interacts with physical dimensions that are approximately the same size as the wavelength of the radiation.

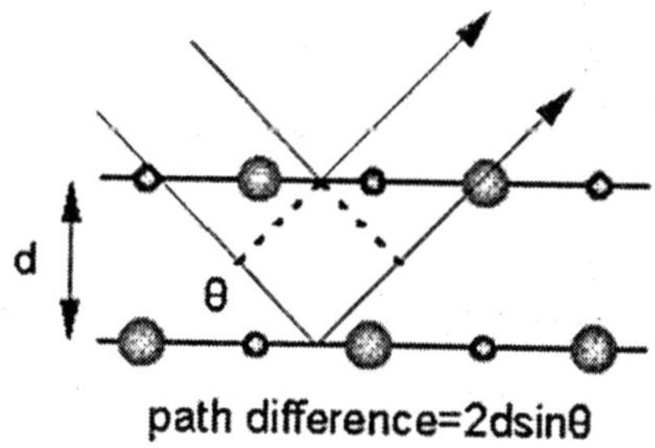

Fig. Interference of radiation between atomic planes in a crystal

These interference effects are useful for determining dimensions in solid materials, and therefore crystal structures. Since the distances between atoms or ions is on the order of 10^{-10} m (1 Å), diffraction methods require radiation in the x-ray region of the electromagnetic spectrum, or beams of electrons or neutrons with a similar wavelength. Electrons and neutrons are commonly thought of as particles, but they have wave properties with the wavelength depending on the energy of the particles as described by the de Broglie equation. The three diffraction methods have different properties that are described in more detail in separate documents. For example, the penetration depths of the three types of beams are quite different:

neutrons > x-rays > electrons.

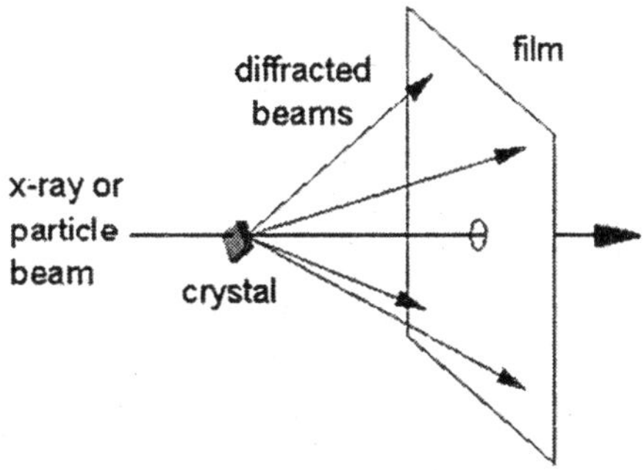

Fig. Schematic of crystal-structure determination by diffraction

ELECTRON DIFFRACTION

Electron diffraction provides similar structural information as neutron diffraction. Electron beams strongly interact with nuclei and electron diffraction is more useful than x-ray diffraction for determining proton positions. The strong interaction of electrons with matter results in a low penetration depth, and electron diffraction is usually used in a reflection geometry to study surfaces or thin films. Electron beams are easy to manipulate, detect, and focus to small spots to provide high spatial resolution.

Electrons scatter from gases and electron diffraction must be performed under vacuum.

NEUTRON DIFFRACTION

Neutron diffraction provides similar structural information as electron diffraction. Neutron beams interact more strongly with nuclei than do x-rays and neutron diffraction is more useful than x-ray diffraction for determining proton positions. Neutrons interact with a solid to a much lesser degree than x-rays and therefore have advantages in studying materials that are damaged by x-rays and in cases where a large penetration depth is desired. For the three types of diffraction methods, neutrons are unique in that they have a magnetic moment and are therefore sensitive to magnetic ordering in a solid.

Neutrons are produced by nuclear reactions in either a nuclear reactor or in an accelerator. Reactor sources produce a continuous spectrum of neutron energies and require a monochromator crystal to select a particular energy. Accelerator sources are usually operated in a pulsed mode and neutron wavelength is selected by time-of-flight methods, that is, data is taken at a fixed Bragg angle as a function of neutron energy.

X-RAY DIFFRACTION (XRD)

The wavelengths of x-rays are of the same order of magnitude as the distances between atoms or ions in a molecule or crystal (Å, 10^{-10} m). A crystal diffracts an x-ray beam passing through it to produce beams at specific angles depending on the x-ray wavelength, the crystal orientation, and the structure of the crystal. X-rays are predominantly diffracted by electron density and analysis of the diffraction angles produces an electron density map of the crystal. Since hydrogen atoms have very little electron density, determining their positions requires extensive refinement of the diffraction pattern. Electron diffraction and neutron diffraction are sensitive to nuclei and are often used to accurately determine hydrogen positions.

X-ray diffractometers consist of an x-ray generator, a goniometer and sample holder, and an x-ray detector such as photographic film or a movable proportional counter. X-ray tubes generate x-rays by bombarding a metal target with high-energy (10 - 100 keV) electrons that knock out core electrons. An electron in an outer shell fills thehole in the inner shell and emits an x-ray photon. Two common targets are Mo and Cu, which have strong K(alpha) x-ray emission at 0.71073 and 1.5418 Å, respectively. X-rays can also be generated by decelerating electrons in a target or a synchrotron ring. These sources produce a continuous spectrum of x-rays and require a crystal monochromator to select a single wavelength.

Fig. Picture of a single-crystal X-ray diffractometer

POWDER X-RAY DIFFRACTION

Powders of crytalline materials diffract x-rays. A beam of x-rays passing through a sample of randomly-oriented microcrystals produces a pattern of rings on a distant screen. Powder x-ray diffraction provides less information than single-crystal diffraction, however, it is much simpler and faster. Powder x-ray diffraction is useful for confirming the identity of a solid material and determining crystallinity and phase purity.

Modern powder x-ray diffractometers consist of an x-ray source, a movable sample platform, an x-ray detector, and associated computer-controlled electronics. The sample is either packed into a shallow cup-shaped holder or deposited as a slurry onto a quatz substrate, and the sample holder spins slowly during the experiment to reduce sample heating. The x-ray source is usually the same as used in single-crystal diffractometers, Mo or Cu. The x-ray beam is fixed and the sample platform rotates with respect to the beam by an angle theta. The detector rotates at twice the rate of the sample and is at an angle of 2theta with respect to the incoming x-ray beam.

IMAGING TECHNIQUES

Microscopy uses radiation and optics to obtain a magnified image of an object. The resolution of the imaging is limited by the minimum focus of the radiation due to diffraction. For light microscopy the diffraction limit is approximately 1 um (10^{-6} m) and for high-resolution electron microscopy the limit is approximately 1 Å (10^{-10} m). Several new imaging techniques use very sharp tips or probes to acheive atomic-scale resolution (AFM, STM), or to obtain optical images below the diffraction limit (NFOM).

SPECIFIC MICROSCOPY TECHNIQUES

- Atomic-force microscopy (AFM)
- Light microscopy
- Near-field optical microscopy (NFOM)
- Scanning electron microscopy (SEM)
- Scanning-tunneling microscopy (STM)
- Transmission electron microscopy (TEM)

ATOMIC-FORCE MICROSCOPY (AFM)

Atomic-force microscopy is similar to scanning-tunneling microscopy (STM) in that it can image surfaces at atomic-scale resolution. The difference between AFM and STM is that AFM does not require that the sample be an electrically conducting material. Like STM it uses an atomically-sharp tip that is brought very close to the surface. The tip will feel a chemical attraction or repulsion and will move up or down on its supporting cantilever. The key to the sensitivity of AFM is in monitoring the movement of the tip. A common means of monitoring the tip movement is to use a laser beam that is reflected or diffracted by the tip or cantilever. Up or down movement of the tip is then detected by changes in the laser beam position. As in STM, rastering the tip across the surface produces a topographic map of the surface with atomic resolution.

LIGHT MICROSCOPY

Light microscopy uses electromagnetic radiation in the ultraviolet or visible wavelength range to obtain a magnified image of an object. The resolution of the imaging is limited by the minimum focus of the radiation due to diffraction. For light microscopy the diffraction limit is approximately 1 um (10^{-6} m). Imaging objects smaller than 1 um is possible using electrons instead of light. High-resolution electron microscopy can image features as small as approximately 1 Å (10^{-10} m).

MICROSCOPES

The simplest light microscope consists of an objective lens and an eyepiece. Microscope objectives and eyepieces usually consist complex lens systems of two or more lenses to correct for lens aberrations.

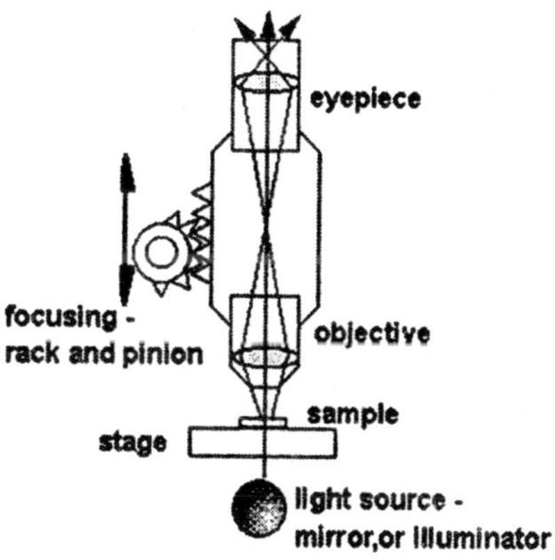

Fig. Schematic of a light microscope

The objective lens forms a real intermediate image which is then greatly magnified by the eyepiece. The objective lens and eyepiece are maintained at a fixed distance and focusing is achieved by moving the whole assembly up and down in relation to the sample. High magnification requires very bright illumination of the sample and a condensor lens is usually placed between the light source and the sample stage to focus light onto the sample.

NEAR-FIELD OPTICAL MICROSCOPY (NFOM)

Light at optical wavelengths, 500 nm for example, cannot be focused to spots much smaller than approximately one micrometer due to diffraction. This diffraction limit prevents optical imaging beyond the micrometer scale. Near-field optical microscopy (NFOM) uses a light source of less than 100-nm diameter to beat the diffraction limit of conventional optical microscopy. The main concept of NFOM is to place the sample very close to the light source, i.e., in the near field, so that the imaging resolution is determined by the diameter of the light source.

The near-field light source is made from a glass capillary that is heated with a CO_2 laser and pulled to an atomically-sharp tip. The outside of the capillary is coated with silver for reflectivity except at the very end of the tip. Laser light is focused into the glass capillary and propagates to the tip by internal reflection (as in optical fibers). A small amount of the light leaks out of the tip via the evanescent wave. The near-field source is brought very close to a sample, and the sample is rastered back and forth to produce an image. NFOM can use either absorption or fluorescence as the optical signal.

ELECTRON MICROSCOPY

Electron microscopy is an imaging technique that uses an electron beam to probe a material. Since the wavelength of electrons is much smaller than the wavelength of visible light, diffraction effects occur at much smaller physical dimensions. The imaging resolution in electron microscopy is therefore much better than in light microscopy.

The use of electron beams requires that the sample be placed in a vacuum chamber for analysis. An electron beam is produced by applying a high voltage to a hot tungsten filament, and accelerating the emitted electrons through a high electric field, 10-50 kV. The electron beam is then focussed with magnetic field lenses to a spot of 100 nm or less on the sample.

A recent innovation in electron microscopy is the use of several stages of differential pumping between the electron gun and the sample, which is placed in a vacuum of a few torr. This method is called environmental electron microscopy and allows imaging of samples that would quickly vapourize in a high-vacuum environment.

TRANSMISSION ELECTRON MICROSCOPY (TEM)

TEM images the electrons that pass through a sample. Since electrons interact strongly with matter, electrons are attenuated as they pass through a solid requiring the samples to be prepared in very thin sections. The image of the sample is observed on a phosphor screen below the sample, and can be recorded with film.

SCANNING ELECTRON MICROSCOPY (SEM)

SEM images the electrons that are reflected from a sample. These images are useful for studying surface morphology or measuring particle sizes. The electron beam is rastered across the sample by ramping voltages on the x- and y-deflection plates through which the electron beam passes (the z axis is the electron-beam direction). Nonconductive samples require an evaporated gold coating over the sample to prevent charging effects that would distort the electric fields in the electron microscope.

SCANNING TUNNELING MICROSCOPY (STM)

Scanning-tunneling microscopy (STM) can image surfaces of conducting materials with atomic-scale resolution. It uses an atomically-sharp metal tip that is brought very close to the surface. When the tip and sample are connected with a voltage source, a small tunneling current flows between the tip and sample surface. This current can be measured, and the magnitude depends on the distance between the tip and the surface. As the tip is moved laterally across the surface, a feedback mechanism moves the tip up and down to maintain a constant tunneling current. Rastering the tip across the surface therefore produces a topographic map of the surface.

WAVE OPTICS

INTRODUCTION

Geometric optics is an incredibly successful theory. Probably its most important application is in describing and explaining the operation of commonly occurring optical instruments: *e.g.*, the camera, the telescope, and the microscope. Although geometric optics does not make any explicit assumption about the nature of light, it tends to suggest that light consists of a stream of massless particles. This is certainly what scientists, including, most notably, Isaac Newton, generally assumed up until about the year 1800.

Let us examine how the particle theory of light accounts for the three basic laws of geometric optics:

- *The law of geometric propagation:* This is easy. Massless particles obviously move in straight-lines in free space.
- *The law of reflection:* This is also fairly easy. We merely have to assume that light particles bounce *elastically* (*i.e.*, without energy loss) off reflecting surfaces.
- *The law of refraction:* This is the tricky one. Let us assume that the speed of light particles propagating through a transparent dielectric medium is proportional to the index of refraction,. Let us further assume that at a general interface between two different dielectric media, light particles crossing the interface conserve momentum in the plane parallel to the interface. In general, this implies that the particle momenta normal to the interface are not conserved: *i.e.*, the interface exerts a normal reaction force on crossing particles, but no parallel force. From Fig. , parallel momentum conservation for light particles crossing the interface yields $n \equiv \sqrt{k}$.

  $\upsilon_1 \sin\theta_1 = \upsilon_2 \sin\theta_2$

- However, by assumption $\upsilon_1 = n_1 c$, and $\upsilon_2 = n_2 c$, so

  $n_1 \sin\theta_1 = n_2 \sin\theta_2$

- This highly contrived (and incorrect) derivation of the law of refraction was first proposed by Descartes in 1637. Note that it depends crucially on the (incorrect) assumption that light travels *faster* in dense media (*e.g.*, glass) than in rarefied media (*e.g.*, water). This assumption appears very strange to us nowadays, but it seemed eminently reasonable to scientists in the 17th and 18th centuries. After all, they knew that sound travels faster in dense media (*e.g.*, water) than in rarefied media (*e.g.*, air).

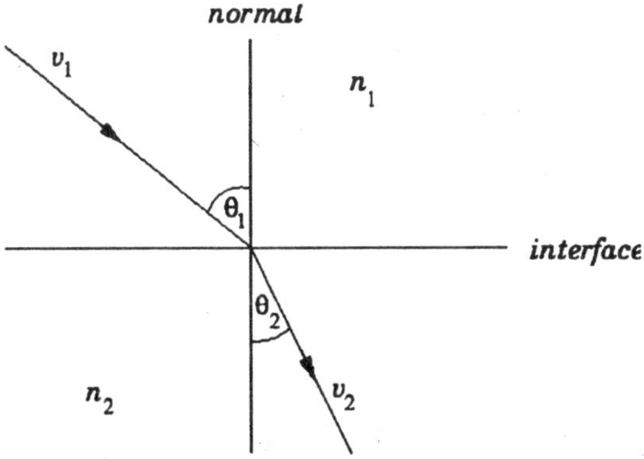

Fig. Descartes' model of refraction

The wave theory of light, which became established in the first half of the 19th century, initially encountered tremendous resistance. Let us briefly examine the reasons why scientists in the early 1800s refused to think of light as a wave phenomenon? Firstly, the particle theory of light was intimately associated with Isaac Newton, so any attack on this theory was considered to be a slight to his memory. Secondly, all of the waves that scientists were familiar with at that time manifestly did not travel in straight-lines. For instance, water waves are diffracted as they pass through the narrow mouth of a harbour, as shown in Fig.. In other words, the "rays" associated with such waves are bent as they traverse the harbour mouth. Scientists thought that if light were a wave phenomenon then it would also not travel in straight-lines: *i.e.*, it would not cast straight, sharp shadows, any more than water waves cast straight, sharp "shadows." Unfortunately, they did not appreciate that if the wavelength of light is much shorter than that of water waves then light can be a wave phenomenon and still propagate in a largely geometric manner.

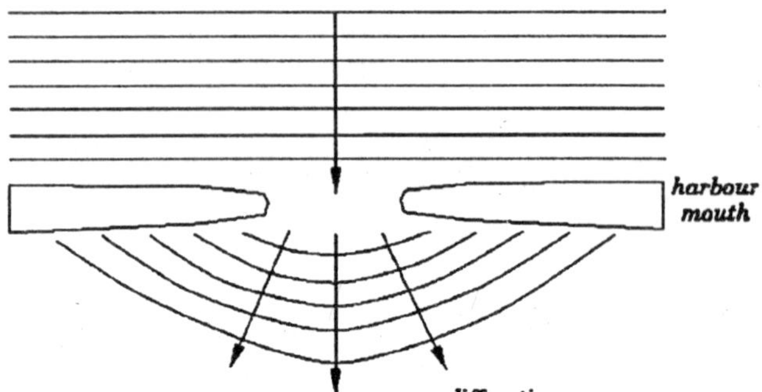

Fig. Refraction of water waves through the entrance of a harbour.

HUYGENS' PRINCIPLE

The first person to explain how wave theory can also account for the laws of geometric optics was Christiaan Huygens in 1670. At the time, of course, nobody took the slightest notice of him. His work was later rediscovered after the eventual triumph of wave theory.

Huygens had a very important insight into the nature of wave propagation which is nowadays called *Huygens' principle.* When applied to the propagation of light waves, this principle states that:

Every point on a wave-front may be considered a source of secondary spherical wavelets which spread out in the forward direction at the speed of light. The new wave-front is the tangential surface to all of these secondary wavelets.

According to Huygens' principle, a plane light wave propagates though free space at the speed of light,. The light rays associated with this wave-front propagate in straight-lines, as shown in Fig. It is also fairly straightforward to account for the laws of reflection and refraction using Huygens' principle.

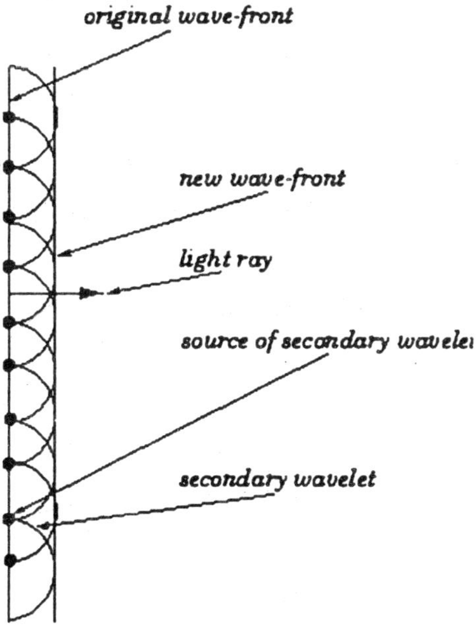

Fig. Huygen's principle.

YOUNG'S DOUBLE-SLIT EXPERIMENT

The first serious challenge to the particle theory of light was made by the English scientist Thomas Young in 1803. Young possessed one of the most brilliant minds in the history of science. A physician by training, he was the first to describe how the lens of the human eye changes shape in order to focus on objects at differing distances. He also studied Physics, and, amongst other things, definitely established the wave theory of light, as described below. Finally, he also studied Egyptology, and helped decipher the Rosetta Stone.

Young knew that sound was a wave phenomenon, and, hence, that if two sound waves of equal intensity, but 180°out of phase, reach the ear then they cancel one another out, and no sound is heard. This phenomenon is called *interference*. Young reasoned that if light were actually a wave phenomenon, as he suspected, then a similar interference effect should occur for light. This line of reasoning lead Young to perform an experiment which is nowadays referred to as

Young's double-slit experiment.

In Young's experiment, two very narrow parallel slits, separated by a distance d, are cut into a thin sheet of metal. Monochromatic light, from a distant light-source, passes through the slits and eventually hits a screen a comparatively large distance L from the slits. The experimental setup is sketched in Fig. .

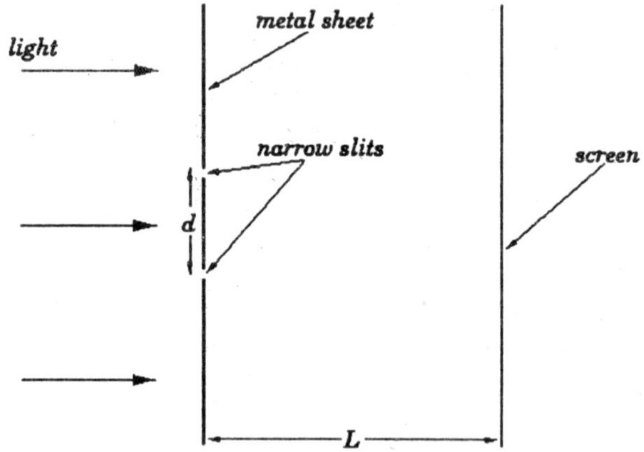

Fig. Young's double-slit experiment.

According to Huygens' principle, each slit radiates spherical light waves. The light waves emanating from each slit are superposed on the screen. If the waves are out of phase then *destructive interference* occurs, resulting in a dark patch on the screen. On the other hand, if the waves are completely in phase then *constructive interference* occurs, resulting in a light patch on the screen.

The point on the screen which lies exactly opposite to the centre point of the two slits, as shown in Fig. is obviously associated with a bright patch. This follows because the path-lengths from each slit to this point are the same. The waves emanating from each slit are initially in phase, since all points on the incident wave-front are in phase (*i.e.*, the wave-front is straight and parallel to the metal sheet). The waves are still in phase at point since they have traveled equal distances in order to reach that point.

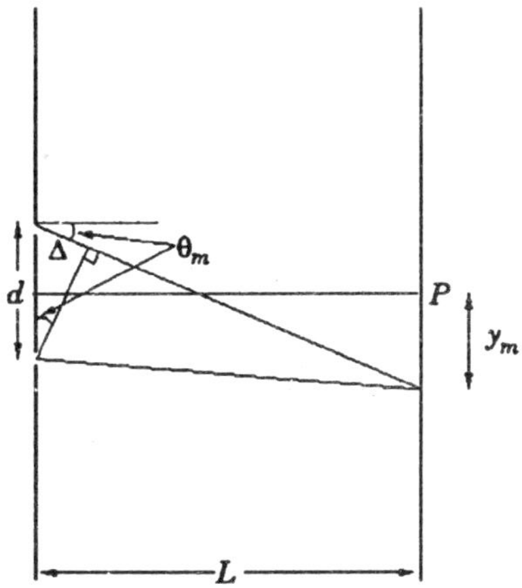

Fig. Interference of light in Young's double-slit experiment.

From the above discussion, the general condition for constructive interference on the screen is simply that the difference in path-length Δ between the two waves be an *integer* number of wavelengths. In other words,

$$\Delta = m\lambda$$

where *m* =0,1,2,.... Of course, the point corresponds to the special case where. It follows, from Fig. that the angular location of the th bright patch on the screen is given by

$$\sin\theta_m = \frac{\Delta}{d} = \frac{m\lambda}{d}$$

Likewise, the general condition for destructive interference on the screen is that the difference in path-length between the two waves be a *half-integer* number of wavelengths. In other words,

$$\Delta = (m + 1/2)\,\lambda;$$

where *m* =0,1,2,.... It follows that the angular coordinate of the th dark patch on the screen is given by

$$\sin\theta'_m = \frac{\Delta}{d} = \frac{(m + 1/2)\,\lambda}{d}.$$

Usually, we expect the wavelength λ of the incident light to be much less than the perpendicular distance L to the screen. Thus,

$$\sin\theta_m \simeq \frac{y_m}{L}:$$

where *ym* measures position on the screen relative to the point *P*.

It is clear that the interference pattern on the screen consists of *alternating light and dark bands,* running parallel to the slits. The distances of the centres of the various light bands from the point *P* are given by

$$y_m = \frac{m\lambda L}{d}:$$

where m = 1,2,3,... Likewise, the distances of the centres of the various dark bands from the point *P* are given by

$$y'_m = \frac{(m+1/2)\lambda L}{d}:$$

where m = 1,2,3,... The bands are *equally spaced,* and of thickness $\lambda L/d$. Note that if the distance from the screen L is much larger than the spacing between the two slits then the thickness of the bands on the screen greatly exceeds the wavelength λ of the light. Thus, given a sufficiently large ratio L/d, it should be possible to observe a banded interference pattern on the screen, despite the fact that the wavelength of visible light is only of order 1 micron. Indeed, when Young performed this experiment in 1803 he observed an interference pattern of the type described above. Of course, this pattern is a *direct consequence* of the wave nature of light, and is completely inexplicable on the basis of geometric optics.

It is interesting to note that when Young first presented his findings to the Royal Society of London he was ridiculed. His work only achieved widespread acceptance when it was confirmed, and greatly extended, by the French physicists Augustin Fresnel and Francois Argo in the 1820s. The particle theory of light was dealt its final death-blow in 1849 when the

French physicists Fizeau and Foucault independently demonstrated that light propagates *more slowly* though water than though air. That the particle theory of light can only account for the law of refraction on the assumption that light propagates *faster* through dense media, such as water, than through rarefied media, such as air.

INTERFERENCE IN THIN FILMS

In everyday life, the interference of light most commonly gives rise to easily observable effects when light impinges on a thin film of some transparent material. For instance, the brilliant colours seen in soap bubbles, in oil films floating on puddles of water, and in the feathers of a peacock's tail, are due to interference of this type.

Suppose that a very thin film of air is trapped between two pieces of glass, as shown in Fig. If monochromatic light (*e.g.*, the yellow light from a sodium lamp) is incident *almost normally* to the film then some of the light is reflected from the interface between the bottom of the upper plate and the air, and some is reflected from the interface between the air and the top of the lower plate. The eye focuses these two parallel light beams at one spot on the retina. The two beams produce either destructive or constructive interference, depending on whether their path difference is equal to an odd or an even number of half-wavelengths, respectively.

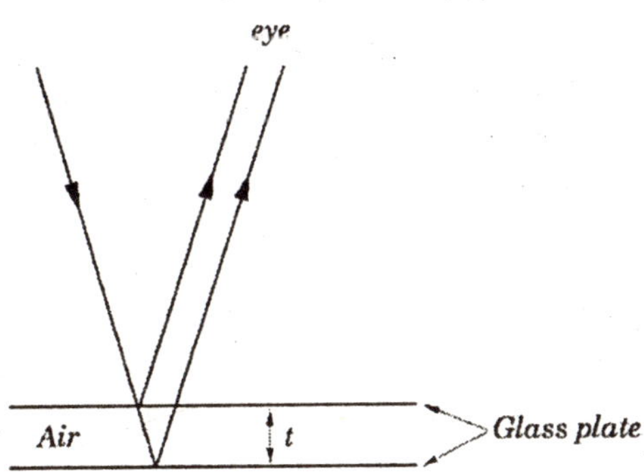

Fig. Interference of light due to a thin film of air trapped between two pieces of glass.

Let be the thickness of the air film. The difference in path-lengths between the two light rays shown in the figure is clearly $\Delta = 2t$. Naively, we might expect that constructive interference, and, hence, *brightness*, would occur if $\Delta = m\lambda$, where is an integer, and destructive interference, and, hence, *darkness*, would occur if $\Delta = (m+1/2)\lambda$. However, this is not the entire picture, since an additional phase difference is introduced between the two rays on reflection. The first ray is reflected at an interface between an optically dense medium (glass), through which the ray travels, and a less dense medium (air). There is no phase change on reflection from such an interface, just as there is no phase change when a wave on a string is reflected from a free end of the string. (Both waves on strings and electromagnetic waves are *transverse waves*, and, therefore, have analogous properties.) The second ray is reflected at an interface between an optically less dense medium (air), through which the ray travels, and a dense medium (glass). There is a 180° phase change on reflection from such an interface, just as there is a 180° phase change when a wave on a string is reflected from a fixed end. Thus, an additional 180° phase change is introduced between the two rays, which is equivalent to an additional path difference of λ/2. When this additional phase change is taken into account, the condition for constructive interference becomes

$$2t = (m + 1/2)\,\lambda;$$

where is an integer. Similarly, the condition for destructive interference becomes

$$2t = m\,\lambda.$$

For white light, the above criteria yield constructive interference for some wavelengths, and destructive interference for others. Thus, the light reflected back from the film exhibits those colours for which the constructive interference occurs.

If the thin film consists of water, oil, or some other transparent material of refractive index then the results are basically the same as those for an air film, except that the wavelength of the light in the film is reduced from λ (the vacuum wavelength) to λ/n. It follows that the modified criteria for constructive and destructive interference are

$$2nt = (m + 1/2)\,\lambda,$$

and

$$2nt = m\,\lambda,$$

respectively.